Student Solutions Manual and Study Guide

for

Serway and Faughn's

College Physics

Seventh Edition
Volume I

John R. Gordon
James Madison University

Charles Teague
Eastern Kentucky University

Raymond A. Serway

THOMSON
BROOKS/COLE

Australia • Canada • Mexico • Singapore • Spain • United Kingdom • United States

Printed in the United States of America
1 2 3 4 5 6 7 09 08 07 06 05

Printer: Thomson/West

ISBN: 0-534-99920-4

For more information about our products,
contact us at:
Thomson Learning Academic Resource Center
1-800-423-0563

For permission to use material from this text or
product, submit a request online at
http://www.thomsonrights.com.
Any additional questions about permissions can be
submitted by email to **thomsonrights@thomson.com.**

Thomson Higher Education
10 Davis Drive
Belmont, CA 94002-3098
USA

Asia (including India)
Thomson Learning
5 Shenton Way
#01-01 UIC Building
Singapore 068808

Australia/New Zealand
Thomson Learning Australia
102 Dodds Street
Southbank, Victoria 3006
Australia

Canada
Thomson Nelson
1120 Birchmount Road
Toronto, Ontario M1K 5G4
Canada

UK/Europe/Middle East/Africa
Thomson Learning
High Holborn House
50–51 Bedford Road
London WC1R 4LR
United Kingdom

Latin America
Thomson Learning
Seneca, 53
Colonia Polanco
11560 Mexico
D.F. Mexico

Spain (including Portugal)
Thomson Paraninfo
Calle Magallanes, 25
28015 Madrid, Spain

Table of Contents

Preface

This *Student Solutions Manual and Study Guide* has been written to accompany the textbook, *College Physics, Seventh Edition*, by Raymond A. Serway and Jerry S. Faughn. The purpose of this ancillary is to provide students with a convenient review of the basic concepts and applications presented in the textbook, together with solutions to selected end-of-chapter problems. This is not an attempt to rewrite the textbook in a condensed fashion. Rather, emphasis is placed upon clarifying typical troublesome points, and providing further practice in methods of problem solving.

Every textbook chapter has a matching chapter in this book and each chapter is divided into several parts. Very often, reference is made to specific equations or figures in the textbook. Each feature of this Study Guide has been included to insure that it serves as a useful supplement to the textbook. Most chapters contain the following components:

- **Notes From Selected Chapter Sections:** This is a summary of important concepts, newly defined physical quantities, and rules governing their behavior.

- **Equations and Concepts:** This is a review of the chapter, with emphasis on highlighting important concepts and describing important equations and formalisms.

- **Suggestions, Skills, and Strategies:** This section offers hints and strategies for solving typical problems that the student will often encounter in the course. In some sections, suggestions are made concerning mathematical skills that are necessary in the analysis of problems.

- **Review Checklist:** This is a list of topics and techniques the student should master after reading the chapter and working the assigned problems.

- **Solutions to Selected End-of-Chapter Problems:** Solutions are given for selected odd-numbered problems that were chosen to illustrate important concepts in each textbook chapter.

- **Tables:** A list of selected Physical Constants is printed on the inside front cover; and a table of some Conversion Factors is provided on the inside back cover.

An important note concerning significant figures: The answers to most of the end-of-chapter problems are stated to three significant figures, though calculations are carried out with as many digits as possible.   We sincerely hope that this *Student Solutions Manual and Study Guide* will be useful to you in reviewing the material presented in the text, and in improving your ability to solve problems and score well on exams. We welcome any comments or suggestions which could help improve the content of this study guide in future editions; and we wish you success in your study.

John R. Gordon
Harrisonburg, VA

Charles Teague
Richmond, KY

Raymond A. Serway
Leesburg, VA

Acknowledgments

We take this opportunity to thank everyone who contributed to this Seventh Edition of *Student Solutions Manual and Study Guide to Accompany College Physics*.

Special thanks for managing and directing this project go to Assistant Editor Annie Mac, Editorial Assistant Brandi Kirksey, Developmental Editor Ed Dodd, Acquisitions Editor Chris Hall, and Publisher for the Physical Sciences David Harris.

Our appreciation goes to our reviewer, Joseph A. Keane of St. Thomas Aquinas College. His careful reading of the manuscript and checking the accuracy of the problem solutions contributed in an important way to the quality of the final product. Any errors remaining in the manual are the responsibility of the authors.

It is a pleasure to acknowledge the excellent work of Martin Arthur of Atelier 88 who prepared the final page layout and camera-ready copy. His technical skills and attention to detail added much to the appearance and usefulness of this volume.

Finally, we express our appreciation to our families for their inspiration, patience, and encouragement.

Suggestions for Study

We have seen a lot of successful physics students. The question, "How should I study this subject?" has no single answer, but we offer some suggestions that may be useful to you.

1. Work to understand the basic concepts and principles before attempting to solve assigned problems. Carefully read the textbook before attending your lecture on that material. Jot down points that are not clear to you, take careful notes in class, and ask questions. Reduce memorization to a minimum. Memorizing sections of a text or derivations does not necessarily mean you understand the material.

2. After reading a chapter, you should be able to define any new quantities that were introduced and discuss the first principles that were used to derive fundamental equations. A review is provided in each chapter of the Study Guide for this purpose, and the marginal notes in the textbook (or the index) will help you locate these topics. You should be able to correctly associate with each physical quantity the symbol used to represent that quantity (including vector notation, if appropriate) and the SI unit in which the quantity is specified. Furthermore, you should be able to express each important principle or equation in a concise and accurate prose statement. Perhaps the best test of your understanding of the material will be your ability to answer questions and solve problems in the text, or those given on exams.

3. Try to solve plenty of the problems at the end of the chapter. The worked examples in the text will serve as a basis for your study. This Study Guide contains detailed solutions to about twelve of the problems at the end of each chapter. You will be able to check the accuracy of your calculations for any odd-numbered problem, since the answers to these are given at the back of the text.

4. Besides what you might expect to learn about physics concepts, a very valuable skill you can take away from your physics course is the ability to solve complicated problems. The way physicists approach complex situations and break them down into manageable pieces is widely useful. Starting in Section 1.10, the textbook develops a general problem-solving strategy that guides you through the steps. To

help you remember the steps of the strategy, they are called *Conceptualize*, *Categorize*, *Analyze*, and *Finalize*.

General Problem-Solving Strategy

Conceptualize

- The first thing to do when approaching a problem is to *think about* and *understand* the situation. Read the problem several times until you are confident you understand what is being asked. Study carefully any diagrams, graphs, tables, or photographs that accompany the problem. Imagine a movie, running in your mind, of what happens in the problem.

- If a diagram is not provided, you should almost always make a quick drawing of the situation. Indicate any known values, perhaps in a table or directly on your sketch.

- Now focus on what algebraic or numerical information is given in the problem. In the problem statement, look for key phrases such as "starts from at rest" ($v_i = 0$), "stops" ($v_f = 0$), or "freely falls" ($a_y = -g = -9.80$ m/s^2). Key words can help simplify the problem.

- Next focus on the expected result of solving the problem. Exactly what is the question asking? Will the final result be numerical or algebraic? If it is numerical, what units will it have? If it is algebraic, what symbols will appear in it?

- Incorporate information from your own experiences and common sense. What should a reasonable answer look like? What should its order of magnitude be? You wouldn't expect to calculate the speed of an automobile to be 5×10^6 m/s.

Categorize

- Once you have a really good idea of what the problem is about, you need to *simplify* the problem. Remove the details that are not important to the solution. For example, you can often model a moving object as a particle. Key words should tell you whether you can ignore air resistance or friction between a sliding object and a surface.

- Once the problem is simplified, it is important to *categorize* the problem. How does it fit into a framework of ideas that you construct to understand the world? Is it a simple *plug-in problem*, such that numbers can be simply substituted into a definition? If so, the problem is likely to be finished when this substitution is done.

If not, you face what we can call an *analysis problem*—the situation must be analyzed more deeply to reach a solution.

- If it is an analysis problem, it needs to be categorized further. Have you seen this type of problem before? Does it fall into the growing list of types of problems that you have solved previously? Being able to classify a problem can make it much easier to lay out a plan to solve it. For example, if your simplification shows that the problem can be treated as a particle moving under constant acceleration and you have already solved such a problem (such as the examples in Section 2.6), the solution to the new problem follows a similar pattern.

Analyze

- Now, you need to analyze the problem and strive for a mathematical solution. Because you have already categorized the problem, it should not be too difficult to select relevant equations that apply to the type of situation in the problem. For example, if your categorization shows that the problem involves a particle moving under constant acceleration, Equations 2.9 to 2.13 are relevant.

- Use algebra (and calculus, if necessary) to solve symbolically for the unknown variable in terms of what is given. Substitute in the appropriate numbers, calculate the result, and round it to the proper number of significant figures.

Finalize

- This final step is the most important part. Examine your numerical answer. Does it have the correct units? Does it meet your expectations from your conceptualization of the problem? What about the algebraic form of the result—before you substituted numerical values? Does it make sense? Try looking at the variables in it to see whether the answer would change in a physically meaningful way if they were drastically increased or decreased or even became zero. Looking at limiting cases to see whether they yield expected values is a very useful way to make sure that you are obtaining reasonable results.

- Think about how this problem compares with others you have done. How was it similar? In what critical ways did it differ? Why was this problem assigned? You should have learned something by doing it. Can you figure out what? Can you use your solution to expand, strengthen, or otherwise improve your framework of ideas? If it is a new category of problem, be sure you understand it so that you can use it as a model for solving future problems in the same category.

When solving complex problems, you may need to identify a series of subproblems and apply the problem-solving strategy to each. For very simple problems, you probably don't need this whole strategy. But when you are looking at a problem and you don't know what to do next, remember the steps in the strategy and use them as a guide.

Work on problems in this Study Guide yourself and compare your solutions with ours. Your solution does not have to look just like the one presented here. A problem can sometimes be solved in different ways, starting from different principles. If you wonder about the validity of an alternative approach, ask your instructor.

5. We suggest that you use this Study Guide to review the material covered in the text, and as a guide in preparing for exams. You can use the sections Chapter Review, Notes From Selected Chapter Sections, and Equations and Concepts to focus in on any points which require further study. The main purpose of this Study Guide is to improve upon the efficiency and effectiveness of your study hours and your overall understanding of physical concepts. However, it should not be regarded as a substitute for your textbook or for individual study and practice in problem solving.

Chapter 1

Introduction

NOTES ON SELECTED CHAPTER SECTIONS

1.1 Standards of Length, Mass, and Time

Systems of units commonly used are the **SI system,** in which the units of mass, length, and time are the kilogram (kg), meter (m), and second (s), respectively; the **cgs** or **gaussian system,** in which the units of mass, length, and time are the gram (g), centimeter (cm), and second, respectively; and the **U.S. customary system**, in which the units of mass, length, and time are the slug, foot (ft), and second, respectively.

The **meter** has been redefined several times. In October 1983, it was redefined to be the distance traveled by light in a vacuum during a time of 1/299 792 458 second.

The SI unit of mass, the **kilogram**, is defined as the mass of a specific platinum-iridium alloy cylinder kept at the International Bureau of Weights and Measures at Sèvres, France.

The **second** is now defined as 9 192 631 700 times the period of one oscillation of radiation from the Cesium-133 atom.

1.2 The Building Blocks of Matter

It is useful to view the atom as a miniature Solar System with a dense, positively charged nucleus occupying the position of the Sun and negatively charged electrons orbiting like the planets. Occupying the nucleus are two basic entities, protons and neutrons. The **proton** is nature's fundamental carrier of positive charge; the **neutron** has no charge and a mass about equal to that of a proton. Even more elementary building blocks than protons and neutrons exist. Protons and neutrons are each now thought to consist of three particles called **quarks**.

1.3 Dimensional Analysis

Dimensional analysis makes use of the fact that dimensions can be treated as algebraic quantities. *Quantities can be added or subtracted only if they have the same dimensions and the terms on each side of an equation must have the same dimensions.*

1.4 Uncertainty in Measurement and Significant Figures

A **significant figure** is a reliably known digit (other than a zero that is used to locate a decimal point).

When **multiplying several quantities**, the number of significant figures in the final result is the same as the number of significant figures in the least accurate of the quantities being multiplied, where "least accurate" means "having the lowest number of significant figures". The same rule applies to division.

When **numbers are added (or subtracted),** the number of decimal places in the result should equal the smallest number of decimal places of any term in the sum (or difference).

Most of the numerical examples and end-of-chapter problems in your textbook will yield answers having either two or three significant figures.

1.5 Conversion of Units

Sometimes it is necessary to **convert units** from one system to another. A list of conversion factors can be found on the inside front cover of the *Student Solutions Manual and Study Guide*.

1.6 Estimates and Order-of-Magnitude Calculations

Often it is useful to **estimate an answer** to a problem in which little information is given. In such a case we refer to the **order of magnitude** of a quantity, by which we mean the power of ten that is closest to the actual value of the quantity. *Usually, when an order-of-magnitude calculation is made, the results are reliable to within a factor of 10.*

1.7 Coordinate Systems

A **coordinate system** used to specify locations in space consists of:

- a fixed reference point O, called the origin

- a set of specified axes, or directions with an appropriate scale and label on each of the axes

- instructions that tell us how to label a point in space relative to the origin and axes

Cartesian (rectangular) coordinates and polar coordinates are two commonly used coordinate systems.

1.8   Trigonometry

You should review the basic trigonometric functions stated by Equations 1.1 and 1.2 in the **Equations and Concepts** section of this chapter.

1.9   Problem-Solving Strategy

In developing problem-solving strategies, the following steps will be helpful.

1. Read the problem carefully at least twice. Be sure you understand the nature of the problem before proceeding further.

2. Draw a suitable diagram with appropriate labels and coordinate axes, if needed.

3. Imagine what happens in the problem.

4. Identify the basic physical principle (or principles) involved, listing the knowns and unknowns.

5. Select a basic equation (or derive an equation) that can be used to find the unknown quantities in terms of the known quantities.

6. Solve the equations for the unknowns symbolically (algebraically without substituting numerical values).

7. Substitute the given values with the appropriate units to obtain numerical values with units for the unknowns.

8. Check your answer against the following questions: Do the units match? Is the answer reasonable? Is the plus or minus sign proper or meaningful?

EQUATIONS AND CONCEPTS

The **three basic trigonometric functions** of one of the acute angles of a right triangle are the sine, cosine, and tangent.

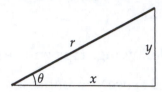

$$\cos \theta = \frac{\text{side adjacent to } \theta}{\text{hypotenuse}} = \frac{x}{r} \qquad (1.1)$$

$$\tan \theta = \frac{\text{side opposite to } \theta}{\text{side adjacent to } \theta} = \frac{y}{x}$$

$$\sin \theta = \frac{\text{side opposite to } \theta}{\text{hypotenuse}} = \frac{x}{y}$$

The **Pythagorean theorem** is an important relationship among the lengths of the sides of a right triangle.

$$r^2 = x^2 + y^2 \qquad (1.2)$$

SUGGESTIONS, SKILLS, AND STRATEGIES

Many mathematical symbols will be used throughout this book. Some important examples are:

$\propto$	denotes	a proportionality
$<$	means	"is less than"
$>$	means	"is greater than"
$<<$	means	"is much less than"
$>>$	means	"is much greater than"
$\cong$	indicates	approximate equality
$=$	indicates	equality
$\sim$	means	"is of the order of"
Δx ("delta x")	indicates	the change in a quantity x
$\lvert x \rvert$	means	the absolute value of x (always positive)
Σ (capital sigma)	represents	a sum. For example,

$$x_1 + x_2 + x_3 + x_4 + x_5 = \sum_{i=1}^{5} x_i$$

REVIEW CHECKLIST

▷ Discuss the units and standards of SI quantities: length, mass and time.

▷ Derive SI units for the quantities force, velocity, volume, acceleration, etc. from units of the three basic quantities: length, mass, and time.

▷ Perform a dimensional analysis of an equation containing physical quantities whose individual units are known.

▷ Convert units from one system to another.

▷ Carry out order-of-magnitude calculations or "guesstimates."

▷ Describe the coordinates of a point in space using both Cartesian and polar coordinate systems.

SOLUTIONS TO SELECTED END-OF-CHAPTER PROBLEMS

4.   Each of the following equations was given by a student during an examination.

$$\tfrac{1}{2}mv^2 = \tfrac{1}{2}mv_0^2 + \sqrt{mgh} \qquad\qquad v = v_0 + at^2 \qquad\qquad ma = v^2$$

Do a dimensional analysis of each equation and explain why the equation can't be correct.

Solution

Any valid equation must be dimensionally correct. That is, all terms in that equation must have the same dimensions as all other terms. However, it is possible for an equation to be dimensionally correct and still be invalid for other reasons.

In the equation $\tfrac{1}{2}mv^2 = \tfrac{1}{2}mv_0^2 + \sqrt{mgh}$, the dimensions of the first two terms are

$$[mv^2] = [mv_0^2] = [m][v^2] = M\left(\frac{L}{T}\right)^2 = \frac{ML^2}{T^2}$$

The dimensions of the third term are

$$[\sqrt{mgh}] = \sqrt{[m][g][h]} = \sqrt{M\left(\frac{L}{T^2}\right)L} = \frac{M^{\frac{1}{2}}L}{T}$$

Since the dimensions of the third term are not the same as those of the first two terms, the equation is dimensionally incorrect and cannot be valid.   ◊

The dimensions of the first two terms in $v = v_0 + at^2$ are

$$[v] = [v_0] = \frac{L}{T}$$

The dimensions of the third term in this equation are

$$[at^2] = [a][t^2] = \left(\frac{L}{T^2}\right)T^2 = L$$

Thus, we see that this equation is dimensionally incorrect and hence invalid.   ◊

In $ma = v^2$, the dimensions of the first term are $[ma] = [m][a] = M\left(\frac{L}{T^2}\right) = \frac{ML}{T^2}$

The dimensions of the last term are $[v^2] = \left(\frac{L}{T}\right)^2 = \frac{L^2}{T^2}$

Again, the equation is seen to be dimensionally incorrect and therefore invalid.   ◊

9. Carry out the following arithmetic operations: (a) the sum of the numbers 756, 37.2, 0.83, and 2.5; (b) the product 0.0032×356.3; (c) the product $5.620 \times \pi$.

Solution

(a) Using a calculator gives a total of 796.53.

In deciding how many significant figures to report in this answer, recall the rule for addition and subtraction:

> "When numbers are added or subtracted, the number of decimal places in the result should equal the smallest number of decimal places of any term in the sum."

Thus, the result should be rounded to 797. ◊

(b) The calculator gives $0.0032 \times 356.3 = (3.2 \times 10^{-3}) \times 356.3 = 1.14016$

The rule for significant figures in products states that

> "When multiplying two or more quantities, the number of significant figures in the final product is the same as the number of significant figures in the least accurate of the quantities multiplied."

Therefore, the product must be rounded to 1.1 because 3.2×10^{-3} has only two significant figures. ◊

(c) The number π is known to a very large number of significant figures, but the product $5.620 \times \pi$ must be rounded to 17.66 because 5.620 has only four significant figures. ◊

12. The radius of a circle is measured to be (10.5 ± 0.2) m. Calculate (a) area and (b) the circumference of the circle and give the uncertainty in each value.

Solution

(a) The area of the circle is

$$A = \pi r^2 = \pi (10.5\,\text{m} \pm 0.2\,\text{m})^2 = \pi (10.5\,\text{m} \pm 0.2\,\text{m})(10.5\,\text{m} \pm 0.2\,\text{m})$$
$$= \pi \left[(10.5\,\text{m})(10.5\,\text{m}) + (10.5\,\text{m})(\pm 0.2\,\text{m}) \right.$$
$$\left. + (\pm 0.2\,\text{m})(10.5\,\text{m}) + (\pm 0.2\,\text{m})(\pm 0.2\,\text{m}) \right]$$
$$= \pi (10.5\,\text{m})^2 \pm 2\pi (10.5\,\text{m})(0.2\,\text{m}) + \pi (0.2\,\text{m})^2$$

Since the last term in this expression is negligible in comparison to the first two terms , the area becomes

$$A = \pi (10.5\,\text{m})^2 \pm 2\pi (10.5\,\text{m})(0.2\,\text{m}) = 346\,\text{m}^2 \pm 13\,\text{m}^2 \qquad \Diamond$$

(b) The circumference of the circle is

$$C = 2\pi r = 2\pi (10.5\,\text{m} \pm 0.2\,\text{m}) = 2\pi (10.5\,\text{m}) \pm 2\pi (0.2\,\text{m})$$

or

$$C = 66.0\,\text{m} \pm 1.3\,\text{m} \qquad \Diamond$$

15. A fathom is a unit of length, usually reserved for measuring the depth of water. A fathom is approximately 6 ft in length. Take the distance from Earth to the Moon to be 250 000 miles, and use the given approximation to find the distance in fathoms.

Solution

To convert the units used to express the distance to the Moon from miles to fathoms, we use a series of ratios in which the numerator and denominator are equal to each other. Each ratio is chosen to move us one step closer to our desired units. Using this technique, we find

$$d = \left(250\,000 \ \cancel{mi}\right)\left(\frac{5\,280 \ \cancel{ft}}{1 \ \cancel{mi}}\right)\left(\frac{1 \ \text{fathom}}{6 \ \cancel{ft}}\right)$$

$$= \frac{(250\,000)(5\,280)}{6} \ \text{fathoms} = 2\times10^8 \ \text{fathoms} \qquad \Diamond$$

Note that the limited accuracy of the given conversion factor from fathoms to feet makes it necessary to round the final answer to one significant figure.

---

21. The speed of light is about 3.00×10^8 m/s. Convert this figure to miles per hour.

Solution

A systematic way to convert units is to multiply the original value by one or more ratios, each ratio chosen so that its value is unity (that is, its numerator and denominator are equal). Further, each ratio should be chosen so that multiplication by it will cancel some of the current units and move you one step closer to the desired units. In the application of this technique given below, notice how the units of the chosen ratios cancel to achieve the requested units.

$$3.00\times10^8 \ \frac{m}{s} = \left(3.00\times10^8 \ \frac{\cancel{m}}{\cancel{s}}\right)\left(\frac{1 \ \cancel{km}}{10^3 \ \cancel{m}}\right)\left(\frac{0.621 \ \text{mi}}{1 \ \cancel{km}}\right)\left(\frac{3\,600 \ \cancel{s}}{1 \ h}\right) = 6.71\times10^8 \ \frac{\text{mi}}{\text{h}} \qquad \Diamond$$

25.   A quart container of ice cream is to be made in the form of a cube. What should be the length of a side, in centimeters? (Use the conversion 1 gallon = 3.786 liter.)

Solution

Since the container is in the form of a cube, with sides of length L, its volume may be written as $V = L \times L \times L = L^3 = 1.00$ quart.

Converting to units of cubic centimeters yields:

$$L^3 = \left(1.00 \ \text{quart}\right)\left(\frac{1 \ \text{gallon}}{4 \ \text{quart}}\right)\left(\frac{3.786 \ \text{liter}}{1 \ \text{gallon}}\right)\left(\frac{1000 \ \text{cm}^3}{1 \ \text{liter}}\right) = 946 \ \text{cm}^3$$

Solving for the length of one side then gives

$$L = \sqrt[3]{946 \ \text{cm}^3} = 9.82 \ \text{cm}$$

◊

---

31.   An automobile tire is rated to last for 50 000 miles. Estimate the number of revolutions the tire will make in its lifetime.

Solution

Tires for full size automobiles often have radii of 15 or 16 inches, or approximately 1.25 ft.

Thus, the circumference of the tire is $C = 2\pi r \approx 8 \ \text{ft}$, and a reasonable guess for the number of revolutions made would be

$$n = \frac{\text{distance traveled}}{\text{circumference}} = \frac{x}{2\pi r} \approx \frac{50\ 000 \ \text{mi}}{8 \ \text{ft/rev}}\left(5\ 280 \ \frac{\text{ft}}{\text{mi}}\right) = 3\times10^7 \ \text{rev} \ ,$$

or $n \sim 10^7 \ \text{rev}$

◊

---

35.   A point is located in a polar coordinate system by the coordinates $r = 2.5$ m and $\theta = 35°$. Find the x and y coordinates of this point, assuming that the two coordinate systems have the same origin.

Solution

The point P is shown in the sketch below with both its polar and Cartesian coordinates labeled. The shaded triangle and basic trigonometry may be used to convert from the polar coordinates to the corresponding Cartesian coordinates:

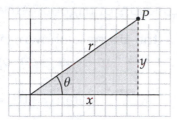

$$\cos\theta = \frac{\text{adjacent side}}{\text{hypotenuse}} = \frac{x}{r} \quad \text{or} \quad x = r\cos\theta$$

$$\sin\theta = \frac{\text{opposite side}}{\text{hypotenuse}} = \frac{y}{r} \quad \text{or} \quad y = r\sin\theta$$

With $r = 2.5$ m and $\theta = 35°$

these yield: $x = 2.0$ m and $y = 1.4$ m   ◊

---

39.   For the triangle shown in Figure P1.39, what are (a) the length of the unknown side, (b) the tangent of θ, and (c) the sine of ϕ?

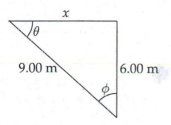

Figure P1.39

Solution

(a)   The unknown side, labeled x in the figure, is most easily found using the Pythagorean Theorem,

$$(9.00 \text{ m})^2 = x^2 + (6.00 \text{ m})^2$$

giving $x^2 = 45.0$ m^2 or $x = 6.71$ m   ◊

(b) $\tan\theta = \dfrac{\text{side opposite } \theta}{\text{side adjacent to } \theta} = \dfrac{6.00 \text{ m}}{x} = \dfrac{6.00 \text{ m}}{6.71 \text{ m}} = 0.894$ ◊

(c) $\sin\phi = \dfrac{\text{side opposite } \phi}{\text{hypotenuse}} = \dfrac{x}{9.00 \text{ m}} = \dfrac{6.71 \text{ m}}{9.00 \text{ m}} = 0.746$ ◊

46.   A surveyor measures the distance across a straight river by the following method: Starting directly across from a tree on the opposite bank, he walks 100 m along the riverbank to establish a baseline. Then he sights across to the tree. The angle from his baseline to the tree is 35.0°. How wide is the river?

Solution

To determine the width w of the river, consider the sketch shown at the right and apply basic concepts of trigonometry.

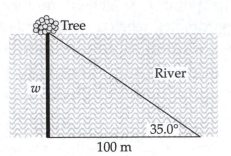

In the triangle shown, notice that the tangent of the 35.0° angle is given by

$$\tan(35.0°) = \frac{\text{opposite side}}{\text{adjacent side}} = \frac{w}{100 \text{ m}}$$

Thus, the width of the river is given by

$$w = (100 \text{ m}) \tan(35.0°) = 70.0 \text{ m} \qquad \Diamond$$

51.   You can obtain a rough estimate of the size of a molecule by the following simple experiment. Let a droplet of oil spread out on a smooth surface of water. The resulting oil slick will be approximately one molecule thick. Given an oil droplet of mass 9.00×10^{-7} kg and density 918 kg/m^3 that spreads out into a circle of radius 41.8 cm on the water surface, what is the order of magnitude of the diameter of an oil molecule?

Solution

The density of an object is its mass divided by its volume. Recognizing this, the volume of oil in the droplet may be found from the equation

$$\text{density} = \frac{\text{mass}}{\text{volume}} : \quad \text{volume} = \frac{\text{mass}}{\text{density}} = \frac{9.00 \times 10^{-7} \text{ kg}}{918 \text{ kg/m}^3} = 9.80 \times 10^{-10} \text{ m}^3$$

The oil slick has a cylindrical shape with a height, h, equal to the diameter of an oil molecule, and a circular cross-section of radius $r = 41.8$ cm $= 0.418$ m. Its volume is then given by

$$\text{volume} = (\text{height})(\text{cross-sectional area})$$

or

$$\text{volume} = h\left(\pi r^2\right) = h\pi (0.418 \text{ m})^2 = h\left(0.549 \text{ m}^2\right)$$

Since the volume of oil in the slick is same as the volume of the droplet,

$$h\left(0.549 \text{ m}^2\right) = 9.80 \times 10^{-10} \text{ m}^3$$

or

$$h = \frac{9.80 \times 10^{-10} \text{ m}^3}{0.549 \text{ m}^2} = 1.78 \times 10^{-9} \text{ m}$$

The order of magnitude of the diameter of an oil molecule is then: $h \sim 10^{-9}$ m   ◊

55.   (a) How many seconds are there in a year? (b) If one micrometeorite (a sphere with a diameter on the order of 10^{-6} m) struck each square meter of the Moon each second, estimate the number of years would it take to cover the Moon with micrometeorites to a depth of one meter. (*Hint:* Consider a cubic box, 1 m on a side, on the Moon, and find how long it would take to fill the box.)

Solution

We use multiple ratios, each with the numerator equal to the denominator, to convert a time interval of 1 year to seconds.

$$1\,\text{yr} = \left(1\,\cancel{\text{yr}}\right)\left(\frac{365.2\,\cancel{\text{day}}}{1\,\cancel{\text{yr}}}\right)\left(\frac{24\,\cancel{\text{h}}}{1\,\cancel{\text{day}}}\right)\left(\frac{60\,\cancel{\text{min}}}{1\,\cancel{\text{h}}}\right)\left(\frac{60\,\text{s}}{1\,\cancel{\text{min}}}\right) = 3.16\times10^{7}\,\text{s} \qquad \Diamond$$

Consider a cubical box, sunk into the Moon, having an open top with an area of $1\,\text{m}^2$ and extending downward to a depth of 1 meter. When filled with micrometeorites, each having a diameter of 10^{-6} m , the number of meteorites along each edge of this box is

$$n = \frac{\text{length of edge}}{\text{meteorite diameter}} = \frac{1\,\text{m}}{10^{-6}\,\text{m}} = 10^{6}$$

The total number of meteorites in the filled box is approximately

$$N \approx n \times n \times n = n^3 = \left(10^6\right)^3 = 10^{18}$$

At the rate of 1 meteorite entering the box each second, the time required to fill the box is

$$t = \frac{N}{\text{rate}} \approx \frac{10^{18}}{1\,\text{per sec}} = \left(10^{18}\,\text{s}\right)\left(\frac{1\,\text{yr}}{3.156\times10^{7}\,\text{s}}\right) = 3\times10^{10}\,\text{yr} \ \text{ or } \ t \sim 10^{10}\,\text{yr} \qquad \Diamond$$

Chapter 2

Motion in One Dimension

NOTES FROM SELECTED CHAPTER SECTIONS

2.1 Displacement

The displacement of an object, defined as its **change in position**, is given by the difference between its final and initial coordinates, or $x_f - x_i$. Displacement is an example of a vector quantity. *A vector is a physical quantity that requires a specification of both direction and magnitude.*

2.2 Velocity

Average speed is the ratio of the total distance traveled to the elapsed time of travel. *Speed is a scalar quantity and is always zero or positive.*

The **average velocity** of an object during the time interval t_i to t_f is equal to the slope of the straight line joining the initial and final points on a graph of the position of the object plotted versus time. *Velocity is a vector quantity and can be positive, negative, or zero.*

Instantaneous velocity (velocity at a specific time) is the slope of the line tangent to the position-time curve at a point P corresponding to the specified time.

The **instantaneous speed** of an object, which is a scalar quantity, is defined as the magnitude of the instantaneous velocity.

2.3 Acceleration

The **average acceleration** during a given time interval is defined as the change in velocity divided by the time interval during which this change occurs.

The **instantaneous acceleration** of an object at a certain time equals the slope of the tangent to the velocity-time graph at that instant of time.

2.4 Motion Diagrams

The **motion of an object** can be represented by a motion diagram with vectors indicating the direction and relative magnitude of the velocity and acceleration at successive time intervals. You should carefully study Active Figure 2.12 in the textbook. This figure illustrates the motion of a car in three different cases:

- Constant positive velocity, zero acceleration.

- Positive velocity, positive acceleration.

- Positive velocity, negative acceleration.

2.5 One-Dimensional Motion with Constant Acceleration

This type of motion is important because it applies to many objects in nature. When an object moves with constant acceleration, the average acceleration equals the instantaneous acceleration. *Equations 2.6 through 2.10 may be used to solve any problem in one-dimensional motion <u>with constant acceleration</u>.*

2.6 Freely Falling Objects

A freely falling body is an object moving freely under the influence of gravity only, regardless of its initial motion.

It is important to emphasize that any freely falling object experiences an acceleration directed downward. *This is true regardless of the direction of motion of the object.* An object thrown upward or downward will experience the same acceleration as an object released from rest. *Once they are in free fall, all objects have an acceleration downward equal to the acceleration due to gravity.*

EQUATIONS AND CONCEPTS

The **displacement** Δx of an object is defined as change in position. Equation 2.1 gives the displacement for a particle moving (in one-dimensional motion) from an initial position x_i to a final position, x_f. *For the case of one-dimensional motion, the usual vector notion is not required.*

$$\Delta x \equiv x_f - x_i$$

The **average velocity** of an object during a time interval is the ratio of the total displacement to the time interval during which the displacement occurred.

$$v_{av} = \frac{\Delta x}{\Delta t} = \frac{x_f - x_i}{t_f - t_i} \tag{2.2}$$

The **instantaneous velocity** v is defined as the limit of the average velocity as the time interval Δt goes to zero. *Note that the instantaneous velocity of an object might have different values from instant to instant.*

$$v \equiv \lim_{\Delta t \to 0} \frac{\Delta x}{\Delta t} \tag{2.3}$$

The **average acceleration** of an object during a time interval is the ratio of the change in velocity to the time interval during which the change in velocity occurs.

$$a_{av} \equiv \frac{\Delta v}{\Delta t} = \frac{v_f - v_i}{t_f - t_i} \tag{2.4}$$

The **instantaneous acceleration** is defined as the limit of the average acceleration as the time interval Δt goes to zero.

$$a \equiv \lim_{\Delta t \to 0} \frac{\Delta v}{\Delta t} \tag{2.5}$$

The **equations of kinematics** can be used to describe one-dimensional motion with constant acceleration along the x axis. Note that each equation shows a different relationship among physical quantities: initial velocity, final velocity, acceleration, time, and displacement. Also in the form that they are shown, it is assumed that $t_i = 0$, $t_f = t$, and for convenience, $\Delta x = x - x_0$. *Remember, Equations 2.6-2.10 are only valid when the acceleration is constant.*

$$v = v_0 + at \tag{2.6}$$

$$v_{av} = \frac{v_0 + v}{2} \tag{2.7}$$

$$\Delta x = \tfrac{1}{2}(v + v_0)t \tag{2.8}$$

$$\Delta x = v_0 t + \tfrac{1}{2}at^2 \tag{2.9}$$

$$v^2 = v_0^2 + 2a\Delta x \tag{2.10}$$

The motion of **an object in free fall** (along the y-axis) can be described by Equations 2.6–2.10 when acceleration, a, is replaced by $-g$ Note that since $a = -g$ in these four equations, the $+y$ axis is predefined to point upward.

$$v = v_0 - gt$$

$$\Delta y = \tfrac{1}{2}(v + v_0)t$$

$$\Delta y = v_0 t - \tfrac{1}{2}gt^2$$

$$v^2 = v_0^2 - 2g\Delta y$$

SUGGESTIONS, SKILLS, AND STRATEGIES

The following procedure is recommended for solving problems involving motion with constant acceleration.

1. Make sure all the units in the problem are consistent. That is, if distances are measured in meters, be sure that velocities have units of m/s and accelerations have units of m/s^2.

2. Choose a coordinate system and make a labeled diagram of the problem including the directions of all displacements, velocities, and accelerations.

3. Make a list of all the quantities given in the problem and a separate list of those to be determined.

4. Select the equations (see Table 2.4) which will enable you to solve for the unknowns.

5. Construct an appropriate motion diagram and check to see if your answers are consistent with the diagram of the problem.

REVIEW CHECKLIST

▷ Define the displacement and average velocity of a particle in motion. Define the instantaneous velocity and understand how this quantity differs from average velocity.

▷ Define average acceleration and instantaneous acceleration.

▷ Construct a graph of displacement versus time (given a function such as $x = 5 + 3t - 2t^2$) for a particle in motion along a straight line. From this graph, you should be able to determine both average and instantaneous values of velocity by calculating the slope of the tangent to the graph.

▷ Apply the motion equations of this chapter (Equations 2.6–2.10) to any situation where the motion occurs under constant acceleration.

▷ Describe what is meant by a body in free fall (one moving under the influence of gravity—where air resistance is neglected). Recognize that the equations of constant accelerated motion apply directly to a freely falling object and that the acceleration is then given by $a = -g$ (where $g = 9.80 \ m/s^2$).

SOLUTIONS TO SELECTED END-OF-CHAPTER PROBLEMS

5. A motorist drives north for 35.0 minutes at 85.0 km/h and then stops for 15.0 minutes. He then continues north, traveling 130 km in 2.00 h. (a) What is his total displacement? (b) What is his average velocity?

Solution

(a) We choose northward to be the positive direction of motion. While the motorist has constant velocity $v = +85.0$ km/h, the average velocity is $v_{av} = v$ and the displacement that occurs is

$$\Delta x_1 = v(\Delta t) = \left(+85.0 \, \frac{km}{h} \right) \left[(35.0 \, min) \left(\frac{1 \, h}{60 \, min} \right) \right] = +49.6 \, km$$

The displacement during the 15.0 min rest stop is $\Delta x_2 = 0$, and the displacement during the final 2.00 h is $\Delta x_3 = +130$ km. Thus, the total displacement for the trip is

$$\Delta x_{total} = \Delta x_1 + \Delta x_2 + \Delta x_3 = 49.6 \, km + 0 + 130 \, km = 180 \, km \qquad \Diamond$$

(b) The duration of the trip is

$$\Delta t_{total} = \Delta t_1 + \Delta t_2 + \Delta t_3 = (35.0 \, min + 15.0 \, min) \left(\frac{1 \, h}{60 \, min} \right) + 2.00 \, h = 2.83 \, h$$

Thus, the average velocity for the trip is

$$v_{av} = \frac{\Delta x_{total}}{\Delta t_{total}} = \frac{+180 \, km}{2.83 \, h} = +63.4 \, km/h \qquad \Diamond$$

11.   A person takes a trip, driving with a constant speed of 89.5 km/h except for a 22.0-min rest stop. If the person's average speed is 77.8 km/h, how much time is spent on the trip and how far does the person travel?

Solution

If t_1 is the time the time the motorist travels at a constant speed of $v_1 = 89.5\,\text{km/h}$, the average speed is $v_{av} = v_1$ and the displacement during this phase of the trip is

$$\Delta x_1 = v_1 t_1 = (89.5\,\text{km/h})t_1$$

The displacement during the 22.0-min rest stop is zero, so the total displacement for the trip is

$$\Delta x_{total} = \Delta x_1 + 0 = (89.5\,\text{km/h})t_1$$

The total time for the trip is: $\Delta t_{total} = t_1 + (22.0\,\text{min})\left(\dfrac{1\,\text{h}}{60\,\text{min}}\right) = t_1 + 0.367\,\text{h}$

Thus, if the average speed for the trip is $v_{av} = 77.8\,\text{km/h}$,

$$\Delta x_{total} = v_{av}\Delta t_{total} \quad \text{becomes} \quad (89.5\,\text{km/h})t_1 = (77.8\,\text{km/h})(t_1 + 0.367\,\text{h})$$

Solving for t_1 yields $t_1 = 2.44\,\text{h}$, so $\Delta t_{total} = 2.44\,\text{h} + 0.367\,\text{h} = 2.80\,\text{h}$ ◊

The total displacement for the trip is then

$$\Delta x_{total} = v_{av}\Delta t_{total} = (77.8\,\text{km/h})(2.80\,\text{h}) = 218\,\text{km}$$ ◊

17.   Find the instantaneous velocities of the tennis player of Figure P2.7 at (a) 0.50 s, (b) 2.0 s, (c) 3.0 s, and (d) 4.5 s.

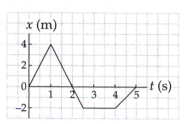

Solution

From Section 2.3 of the textbook, **"The slope of the line tangent to the position-time curve at any point, P, is defined to be the instantaneous velocity at that time."**

Figure P2.7

Thus, we need to determine the slope of the tangent lines to the position-time curve shown in Figure P2.7 at each of the requested times. These slopes may be calculated as follows:

(a) $v_{0.5\,s} = \dfrac{x_{1.0\,s} - x_{0.0\,s}}{1.0\,s - 0.0\,s} = \dfrac{4.0\,m - 0.0\,m}{1.0\,s} = 4.0\,m/s$ ◇

(b) $v_{2.0\,s} = \dfrac{x_{2.5\,s} - x_{1.0\,s}}{2.5\,s - 1.0\,s} = \dfrac{-2.0\,m - 4.0\,m}{1.5\,s} = \dfrac{-6.0\,m}{1.5\,s} = -4.0\,m/s$ ◇

(c) $v_{3.0\,s} = \dfrac{x_{4.0\,s} - x_{2.5\,s}}{4.0\,s - 2.5\,s} = \dfrac{-2.0\,m - (-2.0\,m)}{1.5\,s} = \dfrac{0.0\,m}{1.5\,s} = 0.0\,m/s$ ◇

(d) $v_{4.5\,s} = \dfrac{x_{5.0\,s} - x_{4.0\,s}}{5.0\,s - 4.0\,s} = \dfrac{0.0\,m - (-2.0\,m)}{1.0\,s} = \dfrac{+2.0\,m}{1.0\,s} = 2.0\,m/s$ ◇

21.   A certain car is capable of accelerating at a rate of $+0.60 \text{ m/s}^2$. How long does it take for this car to go from a speed of 55 mi/h to a speed of 60 mi/h?

Solution

The average acceleration over a time interval of duration Δt is defined as

$$a_{av} = \frac{\Delta v}{\Delta t}$$

Thus, the time required to achieve a change in velocity Δv, with an average acceleration a_{av} is

$$\Delta t = \frac{\Delta v}{a_{av}} = \frac{v - v_0}{a_{av}}$$

Since the car maintains a constant acceleration during the time interval of interest, the average acceleration is the same as the constant instantaneous acceleration (that is, $a_{av} = a = +0.60 \text{ m/s}^2$).

The required time is then:

$$\Delta t = \frac{(60 \text{ mi/h} - 55 \text{ mi/h})}{0.60 \text{ m/s}^2} = \left(\frac{5.0 \text{ mi/h}}{0.60 \text{ m/s}^2} \right) \left(\frac{1\,609 \text{ m/mi}}{3\,600 \text{ s/h}} \right) = 3.7 \text{ s} \qquad \Diamond$$

29. A Cessna aircraft has a lift-off speed of 120 km/h. (a) What minimum constant acceleration does the aircraft require if it is to be airborne after a takeoff run of 240 m? (b) How long does it take the aircraft to become airborne?

Solution

Assume the aircraft starts from rest $(v_0 = 0)$ as it accelerates down the runway.

(a) Since the acceleration is uniform, we may use $v^2 = v_0^2 + 2a(\Delta x)$ during the take-off run to find

$$a = \frac{v^2 - v_0^2}{2(\Delta x)} = \frac{\left[(120 \text{ km/h})^2 - 0\right]}{2(240 \text{ m})}\left(\frac{0.278 \text{ m/s}}{1 \text{ km/h}}\right)^2 = 2.32 \text{ m/s}^2 \qquad \lozenge$$

(b) The time required for the aircraft to reach lift-off speed is given by $v = v_0 + at$ as

$$t = \frac{v - v_0}{a} = \frac{(120 \text{ km/h} - 0)}{2.32 \text{ m/s}^2}\left(\frac{0.278 \text{ m/s}}{1 \text{ km/h}}\right) = 14.4 \text{ s} \qquad \lozenge$$

35.   A train is traveling down a straight track at 20 m/s when the engineer applies the brakes, resulting in an acceleration of –1.0 m/s^2 as long as the train is in motion. How far does the train move during a 40-s time interval starting at the instant the brakes are applied?

Solution

It is tempting to try solving this problem by applying the uniformly accelerated motion equation $\Delta x = v_0 t + \dfrac{1}{2} a t^2$ to the full 40 s time interval.

This gives $\Delta x = (20 \ \text{m/s})(40 \ \text{s}) + \dfrac{1}{2}(-1.0 \ \text{m/s}^2)(40 \ \text{s})^2 = 0$

which is an obviously incorrect result.

The source of our error may be found by using $v = v_0 + at$ to find the time required for the train to come to rest.

This yields $t = \dfrac{v - v_0}{a} = \dfrac{0 - 20 \ \text{m/s}}{-1.0 \ \text{m/s}^2} = 20 \ \text{s}$

Therefore, we see that the train will not have a constant acceleration for the full 40-s time interval, so application of uniformly accelerated motion equations to that time interval is invalid. However, the train does have a constant acceleration during the 20-s interval required for it to come to a stop.

Application of $\Delta x = v_0 t + \frac{1}{2} a t^2$ to this interval gives the distance traveled while stopping as

$$\Delta x = (20 \ \text{m/s})(20 \ \text{s}) + \dfrac{1}{2}(-1.0 \ \text{m/s}^2)(20 \ \text{s})^2 = 200 \ \text{m}$$

◊

39.   A hockey player is standing on his skates on a frozen pond when an opposing player, moving with a uniform speed of 12 m/s, skates by with the puck. After 3.0 s, the first player makes up his mind to chase his opponent. If he accelerates uniformly at 4.0 m/s^2, (a) how long does it take him to catch his opponent, and (b) how far has he traveled in that time? (Assume the player with the puck remains in motion at constant speed.)

Solution

(a)   Choose $x = 0$ at the initial location of the player, and $t = 0$ to be the instant when the player starts to chase his opponent. At this time, the opponent is 36 m in front of the player. At time $t > 0$, the displacements of the players from the origin are

$$x_{player} = (x_0)_{player} + (v_0)_{player} t + \frac{1}{2} a_{player} t^2 = 0 + 0 + \frac{1}{2}(4.0 \text{ m/s}^2) t^2 \qquad (1)$$

and, $$x_{opponent} = (x_0)_{opponent} + (v_0)_{opponent} t + \frac{1}{2} a_{opponent} t^2 = 36 \text{ m} + (12 \text{ m/s}) t + 0 \qquad (2)$$

When the players are side-by-side, $x_{player} = x_{opponent}$ $\qquad (3)$

Substituting Equations (1) and (2) into Equation (3) gives

$$\frac{1}{2}(4.0 \text{ m/s}^2) t^2 = (12 \text{ m/s}) t + 36 \text{ m}$$

or $$t^2 - (6.0 \text{ s}) t - 18 \text{ s}^2 = 0$$

This quadratic equation has solutions of $t = -2.2$ s and $t = +8.2$ s. Since the time must be greater than zero, we must choose the positive solution as the time required for the player to overtake his opponent.

$$t = 8.2 \text{ s} \qquad \Diamond$$

(b)   From Equation 1, the displacement the player undergoes while chasing his opponent is

$$x_{player} = \frac{1}{2} a_{player} t^2 = \frac{1}{2}(4.0 \text{ m/s}^2)(8.2 \text{ s})^2 = 1.3 \times 10^2 \text{ m} \qquad \Diamond$$

47.   A small mailbag is released from a helicopter that is descending steadily at 1.50 m/s. After 2.00 s, (a) what is the speed of the mailbag, and (b) how far is it below the helicopter? (c) What are your answers to parts (a) and (b) if the helicopter is rising steadily at 1.50 m/s?

Solution

(a)   If we choose $+y$ to point upwards, and down as the negative direction, the initial velocity of the mailbag is $v_0 = v_h = -1.50$ m/s.

As soon as the bag is released, it becomes a freely falling body with acceleration $a = -g = -9.80$ m/s^2. The velocity of the bag 2.00 s after it was released is:

$$v_{bag} = v_0 + at = -1.50 \text{ m/s} + \left(-9.80 \text{ m/s}^2\right)(2.00 \text{ s}) = -21.1 \text{ m/s}$$

The speed (magnitude of velocity) is: speed $= \left| v_{bag} \right| = 21.1$ m/s   ◊

(b)   After 2.00 s, the bag and the helicopter each have a displacement from the release point of:

$$\Delta y_{bag} = v_0 t + \tfrac{1}{2} at^2 = (-1.50 \text{ m/s})(2.00 \text{ s}) + \tfrac{1}{2}\left(-9.80 \text{ m/s}^2\right)(2.00 \text{ s})^2 = -22.6 \text{ m}$$

$$\Delta y_h = v_h t = (-1.50 \text{ m/s})(2.00 \text{ s}) = -3.00 \text{ m}$$

The difference between them,

$$\Delta y = \Delta y_{bag} - \Delta y_h = \left[-22.6 \text{ m} - (-3.00 \text{ m})\right] = -19.6 \text{ m}$$

so the bag is 19.6 m below the helicopter 2.00 s after it was released.   ◊

(c)   If the helicopter and bag were moving **upward** at the instant of release, then $v_0 = v_h = +1.50$ m/s

Using this value for v_0 in parts (a) and (b), $v_{bag} = -18.1$ m/s

and   speed $= \left| v_{bag} \right| = 18.1$ m/s   ◊

The net displacement of the helicopter and the bag are $\Delta y_{bag} = -16.6$ m and $\Delta y_h = +3.00$ m, with a distance between the two of $\Delta y = \Delta y_{bag} - \Delta y_h = -19.6$ m. Thus, the bag is again 19.6 m below the helicopter 2.00 s after its release.   ◊

51. A student throws a set of keys vertically upward to his fraternity brother, who is in a window 4.00 m above. The brother's outstretched hand catches the keys 1.50 s later. (a) With what initial velocity were the keys thrown? (b) What was the velocity of the keys just before they were caught?

Solution

Taking upward as the positive vertical direction, the keys have a constant acceleration of $a = -9.80 \text{ m/s}^2$ and undergo an upward displacement of $\Delta y = +4.00 \text{ m}$ during the 1.50-s time interval.

(a) Thus, $\Delta y = v_0 t + \dfrac{1}{2} a t^2$ gives the velocity at the beginning of this interval as

$$v_0 = \frac{\Delta y - \frac{1}{2} a t^2}{t} = \frac{4.00 \text{ m} - \frac{1}{2}\left(-9.80 \text{ m/s}^2\right)\left(1.50 \text{ s}\right)^2}{1.50 \text{ s}} = +10.0 \text{ m/s}$$

$v_0 = 10.0 \text{ m/s}$ **upward**   ◊

(b) The velocity of the keys 1.50 s later (that is, just before they are caught) is

$$v = v_0 + a t = +10.0 \text{ m/s} + \left(-9.80 \text{ m/s}^2\right)\left(1.50 \text{ s}\right) = -4.68 \text{ m/s},$$

or $v = 4.68 \text{ m/s}$ **downward**   ◊

57.   A ball is thrown upward from the ground with an initial speed of 25 m/s; at the same instant, another ball is dropped from a building 15 m high. After how long will the balls be at the same height?

Solution

Choose upward as the positive vertical direction. Then, after the balls are released, they are both freely falling objects with acceleration $a = -g$.

During the time interval from the moment the two balls are released until the instant they are at the same height h above the ground, the ball thrown upward undergoes a displacement $\Delta y_1 = y - y_0 = +h - 0$. Thus, $\Delta y = v_0 t + \frac{1}{2}at^2$ gives

$$h = (+25 \text{ m/s})t - \frac{1}{2}gt^2 \tag{1}$$

For this same time interval, the displacement of the dropped ball is $\Delta y_2 = y - y_0 = h - 15 \text{ m}$ and $\Delta y = v_0 t + \frac{1}{2}at^2$ becomes

$$h - 15 \text{ m} = (0)t + \frac{1}{2}(-g)t^2 \text{ or } h = 15 \text{ m} - \frac{1}{2}gt^2 \tag{2}$$

Substituting from Equation 1 for h in Equation 2 yields

$$(25 \text{ m/s})t - \frac{1}{2}\cancel{g}t^2 = 15 \text{ m} - \frac{1}{2}\cancel{g}t^2 \text{ or } t = \frac{15 \text{ m}}{25 \text{ m/s}} = 0.60 \text{ s} \qquad \diamond$$

62. A mountain climber stands at the top of a 50.0-m cliff that overhangs a calm pool of water. She throws two stones vertically downward 1.00 s apart and observes that they cause a single splash. The first stone had an initial velocity of –2.00 m/s. (a) How long after release of the first stone did the two stones hit the water? (b) What initial velocity must the second stone have had, given that they hit the water simultaneously? (c) What was the velocity of each stone at the instant it hit the water?

Solution

Both stones are freely falling objects, with acceleration $a = -g = -9.80 \text{ m/s}^2$, from the instant they leave the thrower's hand until the instant they hit the water. Also, the displacement of each stone during the downward trip is $\Delta y = -50.0 \text{ m}$.

(a) Using $v^2 = v_0^2 + 2a(\Delta y)$, the velocity of the first stone when it reaches the water is found to be

$$v_1 = -\sqrt{(v_0)_1^2 + 2a(\Delta y)} = -\sqrt{(-2.00 \text{ m/s})^2 + 2(-9.80 \text{ m/s}^2)(-50.0 \text{ m})} = -31.4 \text{ m/s}$$

The time of flight of this stone is given by $v = v_0 + at$ as

$$t_1 = \frac{v_1 - (v_0)_1}{a} = \frac{-31.4 \text{ m/s} - (-2.00 \text{ m/s})}{-9.80 \text{ m/s}^2} = 3.00 \text{ s} \qquad \Diamond$$

(b) Since they strike the water simultaneously, the time of flight of the second stone (released 1.00 s after the first stone) is

$$t_2 = t_1 - 1.00 \text{ s} = 3.00 \text{ s} - 1.00 \text{ s} = 2.00 \text{ s}$$

Thus, $\Delta y = v_0 t + \frac{1}{2}at^2$ gives the initial velocity of the second stone as

$$(v_0)_2 = \frac{\Delta y - \frac{1}{2}at_2^2}{t_2} = \frac{-50.0 \text{ m} - \frac{1}{2}(-9.80 \text{ m/s}^2)(2.00 \text{ s})^2}{2.00 \text{ s}} = -15.2 \text{ m/s} \qquad \Diamond$$

(c) From part (a), the final velocity of the first stone is $v_1 = -31.4 \text{ m/s}$ $\qquad \Diamond$

Also, $v_2 = (v_0)_2 + at_2 = -15.2 \text{ m/s} + (-9.80 \text{ m/s}^2)(2.00 \text{ s}) = -34.8 \text{ m/s}$ $\qquad \Diamond$

67.   A stunt man sitting on a tree limb wishes to drop vertically onto a horse galloping under the tree. The constant speed of the horse is 10.0 m/s, and the man is initially 3.00 m above the level of the saddle. (a) What must be the horizontal distance between the saddle and the limb when the man makes his move? (b) How long is he in the air?

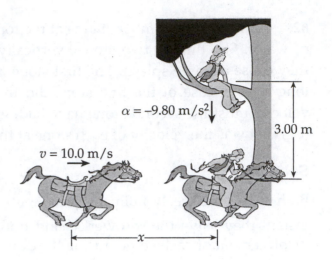

Solution

The most convenient way to solve this problem is to first solve part (b) and use that answer in the solution of part (a) as shown below.

(b)   The stunt man starts from rest $(v_0 = 0)$ and is a freely falling body $(a = -g = -9.80 \text{ m/s}^2)$ until he reaches the saddle. The time to make the 3.00-m vertical drop may be found from $\Delta y = v_0 t + \frac{1}{2} a t^2$.

This gives $\qquad -3.00 \text{ m} = 0 + \frac{1}{2}\left(-9.80 \text{ m/s}^2\right) t^2$

or $\qquad\qquad t = \sqrt{\dfrac{-3.00 \text{ m}}{-4.90 \text{ m/s}^2}} = 0.782 \text{ s}$ ◊

(a)   Because the horse moves with constant velocity, the horizontal distance it travels during the 0.782 s the stunt man is falling (and therefore the horizontal distance that should exist between him and the saddle when he makes his move) is given by:

$$x = v_{\text{horse}} t = (10.0 \text{ m/s})(0.782 \text{ s}) = 7.82 \text{ m}$$ ◊

Chapter 3
Vectors and
Two-Dimensional Motion

NOTES FROM SELECTED CHAPTER SECTIONS

3.1 Vectors and Their Properties

3.2 Components of a Vector

A **scalar** has only magnitude and no direction. On the other hand, a **vector** is a physical quantity that requires the specification of both direction and magnitude. In Cartesian coordinates the **components of a vector** are the projections of the vector along the axes of a rectangular coordinate system. *A vector can be completely described by its components.*

Equal vectors have both the same magnitude and direction. When two or more vectors are **added**, they must have the same units. Two vectors which are the **negative** of each other have the same magnitude but opposite directions. **Multiplication** by a positive (negative) scalar, results is a vector in the same (opposite) direction, with a magnitude equal to the product of the scalar and the magnitude of the original vector.

3.3 Displacement, Velocity, and Acceleration in Two Dimensions

Consider an object moving between two points in space. The **displacement** of the object is defined as the change in the position vector, $\Delta \bar{r}$. The **average velocity** of a particle during the time interval Δt is the ratio of the displacement to the time interval for this displacement. The average velocity is a vector quantity directed along $\Delta \bar{r}$. The **instantaneous velocity** $\bar{v}$ is defined as the limit of the average velocity as Δt goes to zero. The direction of the instantaneous velocity vector is along a line that is tangent to the path of the particle and in the direction of motion. The **average acceleration** of an object whose velocity changes is the ratio of the net change in velocity to the time interval during which the change occurs.

In a given time interval, a particle can accelerate in several ways: the magnitude of the velocity vector (the speed) may change; the direction of the velocity vector may change, making a curved path, even though the speed is constant; or both the magnitude and direction of the velocity vector may change.

3.4   Motion in Two Dimensions

In the case of **projectile motion**, if it is assumed that air resistance is negligible and that the rotation of the Earth does not affect the motion, then:

- the horizontal component of velocity, v_x, remains constant because there is no horizontal component of acceleration

- the vertical component of acceleration is equal to the acceleration due to gravity

- the vertical component of velocity, v_y, and the displacement in the y direction are identical to those of a freely falling body

- projectile motion can be described as a superposition of the two motions in the x and y directions. *The horizontal and vertical components of the motion of a projectile are completely independent of each other.*

Review the procedure recommended in the Suggestions, Skills, and Strategies section for solving projectile motion problems.

3.5   Relative Velocity

Observations made by observers in **different frames of reference** can be related to each other using the techniques of transformation of relative velocities.

When two objects are each moving with respect to a stationary reference frame (e.g. the Earth), each moving object has a **relative velocity** with respect to the other one. You must use vector addition in order to determine the relative velocity of one object with respect to another. See the suggestions for solving relative velocity problems in the Suggestions, Skills, and Strategies section.

EQUATIONS AND CONCEPTS

Vector quantities obey the **commutative law for addition**. In order to add vector $\vec{A}$ to vector $\vec{B}$ using the graphical method, first construct $\vec{A}$ and then draw $\vec{B}$ such that the tail of $\vec{B}$ starts at the head of $\vec{A}$. The sum of $\vec{A} + \vec{B}$ is the vector that completes the triangle by connecting the tail of $\vec{A}$ to the head of $\vec{B}$.

$$\vec{A} + \vec{B} = \vec{B} + \vec{A}.$$

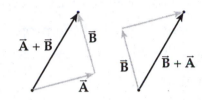

When more than two vectors are to be added by the **graphical or geometric method**, they are all connected head-to-tail in any order and the **resultant or sum** is the vector which joins the tail of the first vector to the head of the last vector. The length of each vector corresponds to the magnitude of the vector according to a chosen scale. *When two or more vectors are to be added, all of them must represent the same physical quantity — that is, have the same units. The direction of each vector must be along a direction which makes the proper angle relative to the others.*

$$\vec{R} = \vec{A} + \vec{B} + \vec{C} + \vec{D}$$

Graphical addition of vectors

The operation of **vector subtraction** utilizes the definition of the negative of a vector. The negative of vector $\vec{A}$ is the vector which has a magnitude equal to the magnitude of $\vec{A}$, but acts or points along a direction opposite the direction of $\vec{A}$.

$$\vec{A} - \vec{B} = \vec{A} + (-\vec{B}) \tag{3.1}$$

Vector subtraction

A vector $\vec{A}$ in a two-dimensional coordinate system can be resolved into its **horizontal and vertical components** along the x and y directions. The projection of $\vec{A}$ onto the x axis, A_x, is the x component of $\vec{A}$; and the projection of $\vec{A}$ onto the y axis, A_y, is the y component of $\vec{A}$.

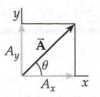

Rectangular components of a vector

The **magnitude of vector** $\vec{A}$ and the angle, θ, which the vector makes with the positive x axis can be determined from the values of the x and y components of $\vec{A}$.

$$A_x = A\cos\theta \qquad (3.2)$$
$$A_y = A\sin\theta$$

$$A = \sqrt{A_x^2 + A_y^2} \qquad (3.3)$$

$$\tan\theta = \frac{A_y}{A_x} \qquad (3.4)$$

The **path of a projectile** is curved as shown in the figure below. Such a curve is called a parabola. The vector which represents the initial velocity, $\vec{v}_0$, makes an angle of θ_0 with the horizontal where θ_0 is called the projection angle or angle of launch. In order to analyze projectile motion, you should separate the motion into two parts, the x (horizontal) motion and the y (vertical) motion, and apply the equations of constant acceleration to each part separately.

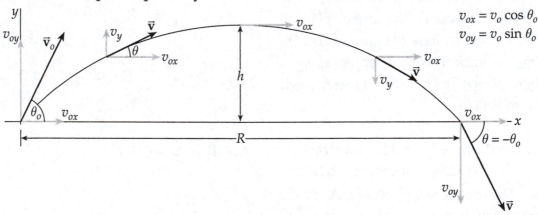

$$v_{ox} = v_o\cos\theta_o$$
$$v_{oy} = v_o\sin\theta_o$$

Velocity components, maximum height, and range in projectile motion

Equations for the **maximum height and range** of a projectile can be found by combining Equations 3.13b and 3.14b to remove t. Note that $\Delta y = h$ when $v_y = 0$ and $\Delta x = R$ when $\Delta y = 0$.

$$h = \frac{v_0^2 \sin^2 \theta_0}{2g}$$

$$R = \frac{v_0^2}{g} \sin(2\theta_0)$$

The **initial horizontal and vertical components of velocity** of a projectile depend on the magnitude of the initial velocity vector and the initial angle of launch.

$$v_{0x} = v_0 \cos \theta_0$$
$$v_{0y} = v_0 \sin \theta_0$$

The **horizontal component** of velocity for a projectile remains constant $(a_x = 0)$; while the **vertical component** decreases uniformly with time $(a_y = -g)$. Note the changing value of v_y and constant value of v_x as shown in the figure on page 34.

$$v_x = v_{0x} = v_0 \cos \theta_0 = \text{constant} \qquad (3.13a)$$

$$v_y = v_0 \sin \theta_0 - gt$$

The **x and y coordinates** of the position of a projectile are functions of the elapsed time. *In Equations 3.13a and 3.14b, the positive direction for the vertical motion is assumed to be upward.*

$$\Delta x = v_{0x}t = (v_0 \cos \theta_0)t \qquad (3.13b)$$

$$\Delta y = (v_0 \sin \theta_0)t - \tfrac{1}{2}gt^2 \qquad (3.14b)$$

The **final velocity** in the y direction can be related to the initial velocity and the displacement as stated in Equation 3.14c.

$$v_y^2 = (v_0 \sin \theta_0)^2 - 2g\Delta y \qquad (3.14c)$$

The **magnitude of the velocity** and the angle that the velocity vector makes with the x-axis at any time can be determined from the values of v_x and v_y.

$$v = \sqrt{v_x^2 + v_y^2}$$

$$\theta = \tan^{-1}\left(\frac{v_y}{v_x}\right)$$

SUGGESTIONS, SKILLS, AND STRATEGIES

ADDITION AND SUBTRACTION OF VECTORS

When two or more vectors are to be added, the following step-by-step procedure is recommended:

1.   Select a coordinate system.

2.   Draw a sketch of the vectors to be added (or subtracted), with a label on each vector.

3.   Find the x and y components of all vectors.

4.   Find the algebraic sum of the x-components of the vectors, and the algebraic sum of the y-components of the vectors. These two sums are the x and y components of the resultant vector.

5.   Use the Pythagorean theorem to find the magnitude of the resultant vector.

6.   Use a suitable trigonometric function (e.g. Equation 3.4) to find the angle the resultant vector makes with the x axis.

SOLVING PROJECTILE MOTION PROBLEMS

The following procedure is recommended for solving projectile motion problems:

1.   Sketch the path of the projectile on a set of coordinate axes. Include the initial velocity vector and the projectile angle.

2.   Resolve the initial velocity vector into x and y components.

3.   Treat the horizontal motion (zero acceleration) and the vertical motion (acceleration $= -g$) independently.

4.   Follow the techniques for solving problems with constant velocity (zero acceleration) to analyze the horizontal motion of the projectile.

5.   Follow the techniques for solving problems with constant acceleration to analyze the vertical motion of the projectile.

SOLVING RELATIVE VELOCITY PROBLEMS

1. Examine the statement of the problem for phrases like "… the velocity of A relative to B…". When a velocity is not explicitly stated as being relative to a specific observer, it is usually relative to the Earth.

2. Show the velocity vectors in a diagram; label each vector involved (usually two) with a letter that reminds you what it represents (for example E for earth).

3. Assemble the three velocities you have identified into a vector equation with subscripts which represent the relationship among the velocities. For example, " the velocity of observer A relative to observer B equals the velocity of observer A relative to the Earth minus the velocity of observer B relative to the Earth" would appear as. $\vec{v}_{AB} = \vec{v}_{AE} - \vec{v}_{BE}$.

4. Solve the equation in step 3 for the unknown velocity. In the case of one-dimensional motion, this will involve one equation and one unknown; for two-dimensional motion there will be two component equations and two unknowns.

REVIEW CHECKLIST

▷ Understand and describe the basic properties of vectors, resolve a vector into its rectangular components, use the rules of vector addition (including graphical solutions for addition of two or more vectors), and determine the magnitude and direction of a vector from its rectangular components.

▷ Sketch a typical trajectory of a particle moving in the xy plane and draw vectors to illustrate the manner in which the displacement, velocity, and acceleration of the particle change with time.

▷ Recognize that two-dimensional motion in the xy plane with constant acceleration is equivalent to two independent motions: constant velocity along the x-direction and constant acceleration along the y-direction.

▷ Recognize the fact that if the initial speed v_o and initial angle θ_o of a projectile motion are known at a given point at $t = 0$, the velocity components and coordinates can be found at any later time t. Furthermore, one can also calculate the horizontal range R and maximum height H if v_o and θ_o are known.

▷ Practice the technique demonstrated in the text to solve relative velocity problems.

SOLUTIONS TO SELECTED END-OF-CHAPTER PROBLEMS

5.   A plane flies from base camp to lake A, a distance of 280 km at a direction of 20.0° north of east. After dropping off supplies, the plane flies to lake B, which is 190 km and 30.0° west of north from lake A. Graphically determine the distance and direction from lake B to the base camp.

Solution

First choose a convenient scale, such as 1 cm = 20 km, and lay out the first two legs of the trip on a scale drawing with the base camp at the origin of the coordinate system. Choose the horizontal axis to be the east-west line and orient the arrow representing the trip to lake A and the arrow representing the trip from lake A to lake B in the specified directions as shown at the right. This process accurately locates the two lakes on your drawing.

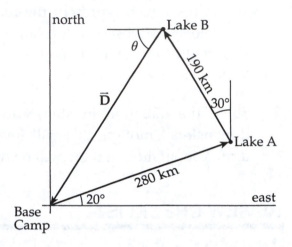

The vector $\vec{D}$ drawn from the location of lake B back to the origin represents the displacement of the base camp from lake B. Measure the length of this vector and multiply by your scale factor to determine the distance you are from camp. Measure the angle θ to determine the direction from lake B back to the base camp. Your results should approximately equal the vector given below:

$$\vec{D} = 310 \text{ km at } 57° \text{ south of west}$$ ◊

11.   A girl delivering newspapers covers her route by traveling 3.00 blocks west, 4.00 blocks north, and then 6.00 blocks east. (a) What is her resultant displacement? (b) What is the total distance she travels?

Solution

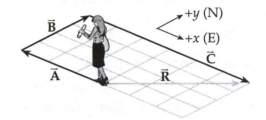

(a)   If eastward and northward are chosen as $+x$ and the $+y$ directions respectively, the components of each displacement are:

Displacement	x-component	y-component
$\vec{A}$	−3.00 blocks	0.00
$\vec{B}$	0.00	+4.00 blocks
$\vec{C}$	+6.00 blocks	0.00

The components of the resultant are then:

$$R_x = A_x + B_x + C_x = +3.00 \text{ blocks}$$

$$R_y = A_y + B_y + C_y = +4.00 \text{ blocks}$$

The magnitude and direction of the resultant are then given by:

$$R = \sqrt{R_x^2 + R_y^2} = 5.00 \text{ blocks} \qquad \diamond$$

$$\theta = \tan^{-1}\left(\frac{R_y}{R_x}\right) = 53.1° \text{ north of east} \qquad \diamond$$

(b)   The distance traveled is $d = (3.00 + 4.00 + 6.00) \text{ blocks} = 13.0 \text{ blocks}$ $\qquad \diamond$

15. The eye of a hurricane passes over Grand Bahama Island in a direction 60.0° north of west with a speed of 41.0 km/h. Three hours later, the course of the hurricane suddenly shifts due north, and its speed slows to 25.0 km/h. How far from Grand Bahama is the hurricane 4.50 h after it passes over the island?

Solution

Assume that Grand Bahama Island lies at the origin of the coordinate system shown in the diagram at the right. During the first three hours after passing this island, the hurricane travels a distance of

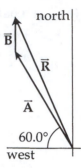

$$A = v_1 \Delta t_1 = (41.0 \ \text{km/h})(3.00 \ \text{h}) = 123 \ \text{km}$$

in a direction 60.0° north of west. Thus, it travels a distance

$$A_{west} = (123 \ \text{km}) \cos 60.0° = 61.5 \ \text{km} \ \ (\text{westward})$$

and

$$A_{north} = (123 \ \text{km}) \sin 60.0° = 106 \ \text{km} \ \ (\text{northward})$$

In the next 1.50 h, the hurricane moves due northward an additional distance of

$$B_{north} = v_2 \Delta t_2 = (25.0 \ \text{km/h})(1.50 \ \text{h}) = 37.5 \ \text{km} \ \ (\text{northward})$$

Thus, 4.50 h after passing Grand Bahama the hurricane is located

$$R_{west} = A_{west} + B_{west} = 61.5 \ \text{km} + 0 = 61.5 \ \text{km west}$$

and

$$R_{north} = A_{north} + B_{north} = 106 \ \text{km} + 37.5 \ \text{km} = 144 \ \text{km north}$$

Its total distance from the island at this time is

$$R = \sqrt{R_{west}^2 + R_{north}^2} = \sqrt{(61.5 \ \text{km})^2 + (144 \ \text{km})^2} = 157 \ \text{km}$$

◊

19. A man pushing a mop across a floor causes the mop to undergo two displacements. The first has a magnitude of 150 cm and makes an angle of 120° with the positive x axis. The resultant displacement has a magnitude of 140 cm and is directed at an angle of 35.0° to the positive x axis. Find the magnitude and direction of the second displacement.

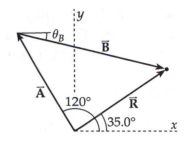

Solution

The vector diagram given above shows the movement of the mop as seen from above the floor. If $\vec{A}$ and $\vec{B}$ are the first and second displacements, the resultant displacement is

$$\vec{R} = \vec{A} + \vec{B}$$

If we take A_x, A_y, B_x, and B_y to be the components of the first and second displacements, the components of the resultant displacement are given by

$$R_x = A_x + B_x \quad \text{and} \quad R_y = A_y + B_y$$

The vectors $\vec{R}$ and $\vec{A}$ are known. Their components are as follows:

x-components:
$$R_x = R\cos\theta_R = (140 \text{ cm})\cos 35.0° = +115 \text{ cm}$$

$$A_x = A\cos\theta_A = (150 \text{ cm})\cos 120° = -75.0 \text{ cm}$$

y-components:
$$R_y = R\sin\theta_R = (140 \text{ cm})\sin 35.0° = +80.3 \text{ cm}$$

$$A_y = A\sin\theta_A = (150 \text{ cm})\sin 120° = +130 \text{ cm}$$

The components of the second displacement $\vec{B}$ may then be found to be:

$$B_x = R_x - A_x = 190 \text{ cm}$$

$$B_y = R_y - A_y = -49.7 \text{ cm}$$

The magnitude of $\vec{B}$ is
$$B = \sqrt{B_x^2 + B_y^2} = \sqrt{(190 \text{ cm})^2 + (-49.7 \text{ cm})^2} = 196 \text{ cm} \qquad \Diamond$$

The direction of $\vec{B}$ is
$$\theta_B = \tan^{-1}\left(\frac{B_y}{B_x}\right) = \tan^{-1}(-0.261) = -14.7° \qquad \Diamond$$

Thus,
$$\vec{B} = 196 \text{ cm at } 14.7° \text{ below the positive } x \text{ direction.} \qquad \Diamond$$

27.   A tennis player standing 12.6 m from the net hits the ball at 3.00° above the horizontal. To clear the net, the ball must rise at least 0.330 m. If the ball just clears the net at the apex of its trajectory, how fast was the ball moving when it left the racquet?

Solution

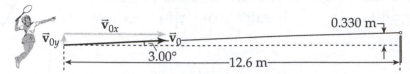

At the apex of the trajectory, $v_y = 0$. Therefore, $v_y = v_{0y} + a_y t$ gives the time to reach the net as

$$t = \frac{v_y - v_{0y}}{a_y} = \frac{0 - v_0 \sin 3.00°}{-g} = \frac{v_0 \sin 3.00°}{g}$$

Since the vertical acceleration is constant, the average velocity in the vertical direction is

$$\left(v_y\right)_{av} = \frac{v_y + v_{0y}}{2}$$

For the time interval from when the ball leaves the racquet until when it reaches the net, this becomes

$$\left(v_y\right)_{av} = \frac{0 + v_0 \sin 3.00°}{2} = \frac{v_0 \sin 3.00°}{2}$$

The vertical displacement during this time interval is

$$\Delta y = \left(v_y\right)_{av} t = \left(\frac{v_0 \sin 3.00°}{2}\right)\left(\frac{v_0 \sin 3.00°}{g}\right) = \frac{v_0^2 \sin^2 3.00°}{2g}$$

If the ball just clears the net, then $\Delta y = 0.330$ m, giving

$$v_0 = \frac{\sqrt{2g(\Delta y)}}{\sin 3.00°} = \frac{\sqrt{2(9.80 \text{ m/s}^2)(0.330 \text{ m})}}{\sin 3.00°} = 48.6 \text{ m/s} \qquad \Diamond$$

Note that it was unnecessary to use the horizontal distance of 12.6 m in this solution.

_____

33.   A projectile is launched with an initial speed of 60.0 m/s at an angle of 30.0° above the horizontal. The projectile lands on a hillside 4.00 s later. Neglect air friction. (a) What is the projectile's velocity at the highest point of its trajectory? (b) What is the straight-line distance from where the projectile was launched to where it hits its target?

Solution

(a)   At the highest point on the trajectory, the projectile's velocity is horizontal with components of

$$v_y = 0 \text{ and } v_x = v_{0x} = v_0 \cos\theta = (60.0 \text{ m/s})\cos 30.0° = 52.0 \text{ m/s}$$

Thus, at the trajectory's highest point, $v = \sqrt{v_x^2 + v_y^2} = 52.0$ m/s

and $\vec{v} = 52.0$ m/s directed horizontally   ◊

(b)   At $t = 4.00$ s after launch, the displacement of the projectile from the launch site has components of

$$\Delta x = v_{0x} t = (v_0 \cos\theta)t = (60.0 \text{ m/s})\cos 30.0°(4.00 \text{ s}) = 208 \text{ m}$$

and

$$\Delta y = v_{0y}t + \tfrac{1}{2}a_y t^2 = (v_0 \sin\theta)t - \tfrac{1}{2}gt^2$$
$$= \left[(60.0 \text{ m/s})\sin 30.0°\right](4.00 \text{ s}) - \tfrac{1}{2}(9.80 \text{ m/s}^2)(4.00 \text{ s})^2 = 41.6 \text{ m}$$

The straight-line distance from the launch site to the landing spot equals the magnitude of the displacement vector, or

$$d = \sqrt{(\Delta x)^2 + (\Delta y)^2} = \sqrt{(208 \text{ m})^2 + (41.6 \text{ m})^2} = 212 \text{ m}$$ ◊

39. A rowboat crosses a river with a velocity of 3.30 mi/h at an angle 62.5° north of west relative to the water. The river is 0.505 mi wide and carries an eastward current of 1.25 mi/h. How far upstream is the boat when it reaches the opposite shore?

Solution

The velocity of the boat relative to the shore, $\vec{v}_{BS}$, may be expressed as the sum $\vec{v}_{BS} = \vec{v}_{BW} + \vec{v}_{WS}$, where $\vec{v}_{BW}$ is the velocity of the boat relative to the water and $\vec{v}_{WS}$ is the velocity of the water relative to the shore.

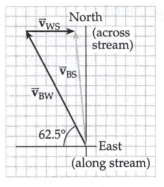

Graphical addition of these vectors is illustrated in the sketch to the right.

Each component of the boat's velocity is as follows:

Directed across the stream (northward),

$$\left(\vec{v}_{BS}\right)_{north} = \left(\vec{v}_{BW}\right)_{north} + \left(\vec{v}_{WS}\right)_{north} = (3.30 \text{ mi/h})\sin 62.5° + 0 = 2.93 \text{ mi/h}$$

Directed parallel to the stream (eastward),

$$\left(\vec{v}_{BS}\right)_{east} = \left(\vec{v}_{BW}\right)_{east} + \left(\vec{v}_{WS}\right)_{east} = -(3.30 \text{ mi/h})\cos 62.5° + 1.25 \text{ mi/h} = -0.274 \text{ mi/h}$$

The time required for the boat to cross the stream (that is, move 0.505 mi north) is given by:

$$t = \frac{0.505 \text{ mi}}{\left(\vec{v}_{BS}\right)_{north}} = \frac{0.505 \text{ mi}}{2.93 \text{ mi/h}} = 0.172 \text{ h}$$

The displacement of the boat parallel to the stream during this time is given by:

$$x = \left(\vec{v}_{BS}\right)_{east} t = (-0.274 \text{ mi/h})(0.172 \text{ h}) = -4.71 \times 10^{-2} \text{ mi}$$

Thus, as the boat crosses the river, it moves in the negative eastward (that is, westward or upstream) direction a distance of

$$|x| = \left(4.71 \times 10^{-2} \text{ mi}\right)\left(\frac{5 \, 280 \text{ ft}}{1.00 \text{ mi}}\right) = 249 \text{ ft}$$

◊

47. Towns A and B in Figure P3.47 are 80.0 km apart. A couple arranges to drive from town A and meet a couple driving from town B at the lake, L. The two couples leave simultaneously and drive for 2.50 h in the directions shown. Car 1 has a speed of 90.0 km/h. If the cars arrive simultaneously at the lake, what is the speed of car 2?

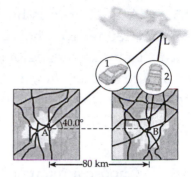
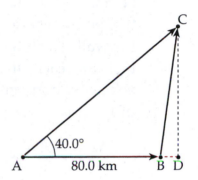

Figure P3.47

Solution

The vector diagram at the right shows the displacement between the cities as $\overrightarrow{\mathbf{AB}}$ (magnitude $\overline{AB} = 80.0$ km), the displacement undergone by Car 1 as $\overrightarrow{\mathbf{AC}}$ (magnitude $\overline{AC}$), and the displacement of Car 2 as $\overrightarrow{\mathbf{BC}}$ (magnitude $\overline{BC}$).

Observe that
$$\overline{AC} = v_1 t = \left(90 \ \frac{\text{km}}{\text{h}}\right)(2.50 \text{ h}) = 225 \text{ km}$$

From the right triangle ADC:
$$\overline{BD} = \overline{AD} - \overline{AB} = \left(\overline{AC}\right)\cos 40.0° - 80.0 \text{ km}$$
$$= (225 \text{ km})\cos 40.0° - 80.0 \text{ km} = 92.4 \text{ km}$$

Finally, using triangle BDC, the Pythagorean theorem gives
$$\overline{BC} = \sqrt{\left(\overline{BD}\right)^2 + \left(\overline{DC}\right)^2} = \sqrt{\left(\overline{BD}\right)^2 + \left(\overline{AC}\sin 40.0°\right)^2}$$
$$\overline{BC} = \sqrt{(92.4 \text{ km})^2 + \left[(225 \text{ km})\sin 40.0°\right]^2} = 172 \text{ km}$$

Hence, the speed of Car 2 is
$$v_2 = \frac{\overline{BC}}{t} = \frac{172 \text{ km}}{2.50 \text{ h}} = 68.6 \text{ km/h}$$

◊

55. A home run is hit in such a way that the baseball just clears a wall 21 m high, located 130 m from home plate. The ball is hit at an angle of 35° to the horizontal, and air resistance is negligible. Find (a) the initial speed of the ball, (b) the time it takes the ball to reach the wall, and (c) the velocity components and the speed of the ball when it reaches the wall. (Assume that the ball is hit at a height of 1.0 m above the ground.)

Solution

(a)   Choose a reference frame with the origin at the point where the ball leaves the bat. The x axis should be horizontal and directed toward the point where the ball crosses the wall. The time required for the ball to reach the wall (that is, achieve a horizontal displacement of 130 m) is

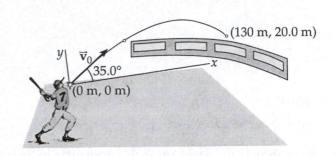

$$t = \frac{\Delta x}{v_{0x}} = \frac{130 \text{ m}}{v_0 \cos 35°} = \frac{159 \text{ m}}{v_0}$$

At this time, the ball must be 21 m above the ground, or 20 m above its launch point ($\Delta y = +20$ m).

Therefore, $\Delta y = v_{0y}t + \frac{1}{2}a_y t^2$

becomes $20 \text{ m} = (v_0 \sin 35°)\left(\frac{159 \text{ m}}{v_0}\right) + \frac{1}{2}(-9.80 \text{ m/s}^2)\left(\frac{159 \text{ m}}{v_0}\right)^2$

Simplifying and solving for the initial velocity gives $v_0 = 42$ m/s   ◊

(b)   From above, the elapsed time when the ball reaches the wall is

$$t = \frac{159 \text{ m}}{v_0} = \frac{159 \text{ m}}{42 \text{ m/s}} = 3.8 \text{ s}$$ ◊

(c)   At this time, the velocity components of the ball are

$$v_x = v_{0x} = v_0 \cos 35° = (42 \text{ m/s})\cos 35° = 34 \text{ m/s},$$ ◊

and $v_y = v_{0y} + a_y t = (42 \text{ m/s})\sin 35° + (-9.80 \text{ m/s}^2)(3.8 \text{ s}) = -13 \text{ m/s}$ ◊

The speed of the ball as it crosses the wall is $v = \sqrt{v_x^2 + v_y^2} = 37$ m/s   ◊

61. By throwing a ball at an angle of 45°, a girl can throw the ball a maximum horizontal distance of R on a level field. How far can she throw the same ball vertically upward? Assume that her muscles give the ball the same speed in each case. (Is this assumption valid?)

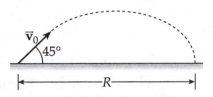

Solution

Throwing the ball at an angle of 45° achieves maximum range for a given initial speed. At this projection angle, the initial velocity components are

$$v_{0x} = v_{0y} = v_0 \sin 45° = v_0 / \sqrt{2}$$

We realize that when the ball returns to ground level, it will have $v_y = -v_{0y}$ and make use of $v_y = v_{0y} + a_y t$ to determine the time of flight as

$$t = \frac{v_y - v_{0y}}{a_y} = \frac{-v_{0y} - v_{0y}}{-g} = \frac{2}{g}\left(\frac{v_0}{\sqrt{2}}\right) = \frac{v_0 \sqrt{2}}{g}$$

The maximum horizontal range is then

$$R = v_{0x}t = \left(\frac{v_0}{\sqrt{2}}\right)\left(\frac{v_0\sqrt{2}}{g}\right) = \frac{v_0^2}{g} \qquad\qquad [1]$$

When the ball is thrown straight upward at speed v_0, the velocity at maximum height is $v_y = 0$ and $v_y^2 = v_{0y}^2 + 2a_y \Delta y$ gives the maximum height reached as

$$\left(\Delta y\right)_{max} = \frac{v_y^2 - v_{0y}^2}{2a_y} = \frac{0 - v_0^2}{-2g} = \frac{v_0^2}{2g}$$

Comparing this result to Equation (1) shows that for a given initial speed,

$$\left(\Delta y\right)_{max} = \frac{R}{2} \qquad\qquad ◊$$

If the girl takes a step when she makes the horizontal throw, she can likely give a higher initial speed to that throw than for the vertical throw.   ◊

65.   A daredevil is shot out of a cannon at 45.0° to the horizontal with an initial speed of 25.0 m/s. A net is positioned a horizontal distance of 50.0 m from the cannon. At what height above the cannon should the net be placed in order to catch the daredevil?

Solution

Choose a reference frame with its origin at the point where the daredevil leaves the cannon, with the x-axis horizontal and the y-axis vertical.

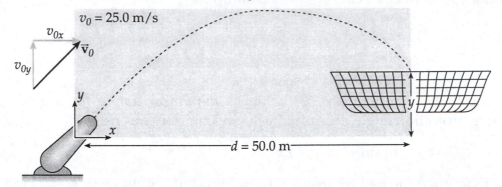

Then, the components of the daredevil's initial velocity and acceleration are:

$$v_{0x} = (25.0 \text{ m/s}) \cos 45° = 17.7 \text{ m/s} \qquad\qquad a_x = 0$$

$$v_{0y} = (25.0 \text{ m/s}) \sin 45° = 17.7 \text{ m/s} \qquad \text{and} \qquad a_y = -g = -9.80 \text{ m/s}^2$$

Thus, the horizontal velocity $v_x = v_{0x} + a_x t$ is constant and the time to travel the horizontal distance of 50.0 m to the net is

$$t = \frac{d}{v_x} = \frac{50.0 \text{ m}}{17.7 \text{ m/s}} = 2.83 \text{ s}$$

The daredevil's y-coordinate at this time is given by $y = v_{0y} t + \frac{1}{2} a_y t^2$ as:

$$y = (17.7 \text{ m/s})(2.83 \text{ s}) + \frac{1}{2}(-9.80 \text{ m/s}^2)(2.83 \text{ s})^2 = +10.8 \text{ m}$$

The net should be placed 10.8 m above the level where the daredevil leaves the cannon.   ◊

71.   One strategy in a snowball fight is to throw a snowball at a high angle over level ground. Then, while your opponent is watching that snowball, you throw a second one at a low angle timed to arrive before or at the same time as the first one. Assume that both snowballs are thrown with a speed of 25.0 m/s. The first is thrown at an angle of 70.0° with respect to the horizontal. (a) At what angle should the second snowball be thrown to arrive at the same point as the first? (b) How many seconds later should the second snowball be thrown after in order for both the first to arrive at the same time?

Solution

(a)   From $\Delta y = v_{0y}t + \frac{1}{2}a_y t^2$, the time (other than $t = 0$) when the first snowball has zero vertical displacement ($\Delta y = 0$) is found to be

$$t_1 = \frac{-2v_{0y}}{a_y} = \frac{-2(25.0 \text{ m/s})\sin 70.0°}{-9.80 \text{ m/s}^2} = 4.79 \text{ s}$$

The horizontal range of this snowball is

$$R_1 = v_{0x}t_1 = \left[(25.0 \text{ m/s})\cos 70.0°\right](4.79 \text{ s}) = 41.0 \text{ m}$$

The time t_2 after its launch when the second snowball, thrown at angle θ with initial speed $v_0 = 25.0$ m/s, returns to ground level is given by $\Delta y = v_{0y}t + \frac{1}{2}a_y t^2$ with $\Delta y = 0$ as

$$t_2 = \frac{-2v_{0y}}{a_y} = \frac{-2(25.0 \text{ m/s})\sin\theta}{-9.80 \text{ m/s}^2} = (5.10 \text{ s})\sin\theta$$

Requiring the horizontal range of this snowball be the same as that of the first ball (that is, $R_2 = v_{0x}t_2 = R_1$) yields

$$\left[(25.0 \text{ m/s})\cos\theta\right]\left[(5.10 \text{ s})\sin\theta\right] = 41.0 \text{ m} \quad \text{or} \quad \sin\theta\cos\theta = 0.321$$

Using the trigonometric identity $\sin 2\theta = 2\sin\theta\cos\theta$ gives

$$\sin 2\theta = 2(0.321) = 0.642 \qquad \text{and} \qquad \theta = 20.0° \qquad \diamond$$

(b)   From part (a) above, the time of flight for the first snowball is $t_1 = 4.79$ s and that for the second snowball is $t_2 = (5.10 \text{ s})\sin\theta = (5.10 \text{ s})\sin 20.0° = 1.74 \text{ s}$.

Thus, if they are to arrive simultaneously, the time delay between the first and second snowballs should be

$$\Delta t = t_1 - t_2 = 4.79 \text{ s} - 1.74 \text{ s} = 3.05 \text{ s} \qquad \diamond$$

Chapter 4
Laws of Motion

NOTES FROM SELECTED CHAPTER SECTIONS

4.1 Forces

Equilibrium is the condition under which the net force (vector sum of all forces) acting on an object is zero. An object in equilibrium has a zero acceleration (velocity is constant or equals zero).

Fundamental forces in nature are:

- gravitational (attractive forces between objects due to their masses)

- electromagnetic forces (between electric charges at rest or in motion)

- strong nuclear forces (between subatomic particles)

- weak nuclear forces (accompanying the process of radioactive decay).

Classical physics is concerned with **contact forces** (which are the result of physical contact between two or more objects) and **action-at-a-distance forces** (which act through empty space and do not involve physical contact).

4.2 Newton's First Law

Newton's first law is called the **law of inertia** and states that an object at rest will remain at rest and an object in motion will remain in motion with a constant velocity unless acted on by a net external force.

Mass and **weight** are two different physical quantities. The weight of a body is equal to the force of gravity acting on the body and varies with location in the Earth's gravitational field. Mass is an inherent property of a body and is a measure of the body's inertia (resistance to change in its state of motion). The SI unit of mass is the **kilogram (kg)** and the unit of weight is the **newton. (N)**

4.3   Newton's Second Law

Newton's second law, the **law of acceleration**, states that the acceleration of an object is directly proportional to the resultant force acting on it and inversely proportional to its mass. *The direction of the acceleration is in the direction of the net force.*

4.4   Newton's Third Law

Newton's third law, the **law of action-reaction**, states that when two bodies interact, the force which body "A" exerts on body "B" (the **action force**) is equal in magnitude and opposite in direction to the force which body "B" exerts on body "A" (the **reaction force**). A consequence of the third law is that forces occur in pairs. *Remember that the action force and the reaction force never cancel because they act on different objects.*

4.5   Applications of Newton's Laws

Construction of a **free-body diagram** is an important step in the application of Newton's laws of motion to solve problems involving bodies in equilibrium or accelerating under the action of external forces. The diagram should **include a labeled arrow to identify each of the external forces** acting on the body whose motion (or condition of equilibrium) is to be studied. *Forces which are the reactions to external forces must not be included.* When a system consists of more than one body or mass, you must construct a free-body diagram for each mass.

4.6   Forces of Friction

When a body is in motion either on a surface or through a viscous medium such as air or water, there is resistance to the motion because the body interacts with its surroundings. We call such resistance a force of friction. Experiments show that the frictional force arises from the nature of the two surfaces. To a good approximation, both $f_{s,\max}$ (maximum force of static friction) and f_k (force of kinetic friction) are proportional to the normal force at the interface between the two surfaces.

EQUATIONS AND CONCEPTS

A **quantitative measurement of mass** (the term used to measure inertia) can be made by comparing the accelerations that a given force will produce on different bodies. If a given force acting on a body of mass m_1 produces an acceleration a_1 and the same force acting on a body of mass m_2 produces an acceleration a_2, the ratio of the two masses equals the inverse of the ratio of the two accelerations.

$$\frac{m_1}{m_2} = \frac{a_2}{a_1}$$

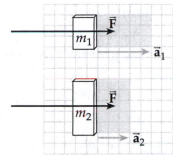

The **acceleration of an object** is proportional to the resultant force acting on it and inversely proportional to its mass. This is a statement of Newton's second law.

$$\sum \vec{F} = m\vec{a} \qquad (4.1)$$

Component scalar equations are equivalent to the vector form of the equation expressing Newton's second law. *The orientation of the coordinate system can often be chosen so that the object has a nonzero acceleration along only one direction.*

$$\sum F_x = m a_x \qquad (4.2)$$
$$\sum F_y = m a_y$$
$$\sum F_z = m a_z$$

The **SI unit of force** is the newton (N), defined as the force that, when acting on a 1-kg mass, produces an acceleration of $1\,\text{m/s}^2$. In the U.S. customary system the unit of force is the pound. *Calculations with Equations 4.1 and 4.2 must be made using a consistent set of units for the quantities force, mass, and acceleration.*

$$1\,\text{N} \equiv 1\,\text{kg}\cdot\text{m/s}^2 \qquad (4.3)$$
$$1\,\text{N} \equiv 0.225\,\text{lb}$$

Newton's **law of universal gravitation** states that every particle in the universe attracts every other particle with a force that is directly proportional to the product of the masses of the particles and inversely proportional to the square of the distance between them.

$$F_g = G\frac{m_1 m_2}{r^2} \tag{4.5}$$

The **universal gravitational constant.**

$$G = 6.67 \times 10^{-11} \text{ N} \cdot \text{m}^2/\text{kg}^2$$

Weight is not an inherent property of a body, but depends on the local value of g and varies with location.

$$w = mg \tag{4.6}$$

The **acceleration due to gravity**, g, decreases with increasing distance from the center of the Earth.

$$g = G\frac{M_E}{r^2} \tag{4.8}$$

Newton's third law, states that forces always occur in pairs; and the force $\vec{F}_{12}$ exerted by body 1 on body 2 (the action force) is equal in magnitude and opposite in direction to the force $\vec{F}_{21}$ exerted by body 2 on body 1 (the reaction force). *The action and reaction forces always act on different objects; they cannot cancel.*

$$\vec{F}_{12} = -\vec{F}_{21}$$

Equilibrium is a condition of rest or motion with constant velocity (magnitude and direction). The vector sum of all forces acting on an object in equilibrium is zero.

$$\sum \vec{F} = 0 \tag{4.9}$$

In the case of **two-dimensional equilibrium** the sum of all forces in the x and y directions must separately equal zero.

$$\sum F_x = 0 \quad \text{and} \quad \sum F_y = 0 \tag{4.10}$$

The **force of static friction** on two objects whose surfaces are in contact but not in motion relative to each other, cannot be greater than $\mu_s n$, where n is the magnitude of the normal (perpendicular) force between the two surfaces and μ_s is a dimensionless constant which depends on the nature of the pair of surfaces.

$$f_s \leq \mu_s n \qquad (4.11)$$

μ_s = coefficient of static friction

A **force of kinetic friction** arises when two surfaces are in relative motion. The force on each body is directed opposite the direction of motion of that body. *The coefficient of kinetic friction μ_k depends on the nature of the two surfaces.*

$$f_k = \mu_k n \qquad (4.12)$$

μ_k = coefficient of kinetic friction

SUGGESTIONS, SKILLS, AND STRATEGIES

For problems involving objects in equilibrium,

1. Make a sketch of the situation described in the problem statement.

2. Draw a free-body diagram for the isolated object under consideration and label all external forces acting on the object. Try to guess the correct direction for each force. If you select a direction that leads to a negative sign in your solution for a force, do not be alarmed; this merely means that the direction of the force is the opposite of what you assumed.

3. Establish a coordinate system and resolve all forces into x and y components.

4. Use the equations $\sum F_x = 0$ and $\sum F_y = 0$ (for objects in equilibrium, acceleration equals zero). Remember to keep track of the signs of the various force components.

5. Application of Step 4 above leads to a set of equations with several unknowns. Solve the simultaneous equations for the unknowns in terms of the known quantities.

For problems involving the application of Newton's second law:

1. Draw a simple, neat diagram of the system and indicate the forces with arrows. Label each force with a symbol representing the nature of the force..

2. Isolate the object of interest whose motion is being analyzed. Draw a free-body diagram for this object; that is, a diagram showing all external forces acting on the object. **For systems containing more than one object, draw a separate diagram for each object. Do not include forces that the object of interest exerts on other objects.**

3. Establish convenient coordinate axes for each object and find the components of the forces along these axes. It is usually convenient to choose the coordinate system so that one of the axes is parallel to the direction of motion of the object.

4. Apply Newton's second law in component form, $(\sum F_x = -ma_x$ and $\sum F_y = -ma_y)$ for each object under consideration.

5. Solve the component equations for the unknowns. *Remember that you must have as many independent equations as you have unknowns in order to obtain a complete solution.* Often in solving such problems, one must also use the equations of kinematics (motion with constant acceleration) from Chapter 2 to find all the unknowns.

REVIEW CHECKLIST

▷   State in your own words a description of Newton's laws of motion; recall physical examples of each law, and identify the action-reaction force pairs in a multiple-body interaction problem as specified by Newton's third law.

▷   Express the normal force in terms of other forces acting on an object, and write out the equation which relates the coefficient of friction, force of friction and normal force between an object and the surface on which it rests or moves.

▷   Apply Newton's laws of motion to various mechanical systems using the recommended procedure discussed in Section 4.5. Most importantly, you should identify all external forces acting on the object of interest, draw the **correct** free-body diagrams which apply to each body of the system, and apply Newton's second law, $\vec{F} = m\vec{a}$, in **component** form.

▷ Apply the equations of kinematics (which involve the quantities: displacement, velocity, and acceleration) as described in Chapter 2 along with those methods and equations of Chapter 4 (involving mass, force, and acceleration) to the solutions of problems where both the kinematic and dynamic aspects are present.

▷ Be familiar with methods for solving two or more linear equations simultaneously for the unknown quantities. Recall that you must have as many independent equations as you have unknowns.

SOLUTIONS TO SELECTED END-OF-CHAPTER PROBLEMS

11.   A boat moves through the water with two forces acting on it. One is a 2 000-N forward push by the water on the propeller, and the other is a 1 800-N resistive force due to the water around the bow. (a) What is the acceleration of the 1 000-kg boat? (b) If it starts from rest, how far will the boat move in 10.0 s? (c) What will its velocity be at the end of that time?

Solution

A resistive force is always directed opposite to the motion. The forward push of the water on the boat's propeller is in the direction of the motion, so the horizontal forces acting on the boat are as shown in the sketch at the right.

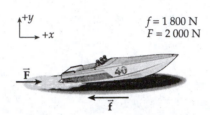

(a)   Taking the forward direction as positive, application of Newton's second law gives the horizontal acceleration as

$$a_x = \frac{\Sigma F_x}{m} = \frac{+2\,000\text{ N} - 1\,800\text{ N}}{1\,000\text{ kg}} = +0.200\text{ m/s}^2 \qquad \Diamond$$

(b)   The horizontal displacement of the boat during the first 10.0 s after it starts from rest is

$$\Delta x = v_{0x}t + \tfrac{1}{2}a_x t^2 = 0 + \tfrac{1}{2}\left(0.200\text{ m/s}^2\right)\left(10.0\text{ s}\right)^2 = 10.0\text{ m} \qquad \Diamond$$

(c)   At the end of this 10.0-s time interval, the horizontal speed of the boat is

$$v_x = v_{0x} + a_x t = 0 + \left(0.200\text{ m/s}^2\right)\left(10.0\text{ s}\right) = 2.00\text{ m/s} \qquad \Diamond$$

14. The force exerted by the wind on the sails of a sailboat is 390 N north. The water exerts a force of 180 N east. If the boat (including its crew) has a mass of 270 kg, what are the magnitude and direction of its acceleration?

Solution

We choose east to be the positive x direction and north the positive y direction. The horizontal forces acting on the 270-kg boat (including crew) are shown at the right.

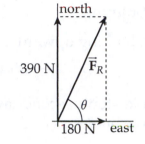

One method of finding the desired acceleration is to first compute the resultant force $\vec{F}_R$:

$$F_R = \sqrt{\left(\Sigma F_x\right)^2 + \left(\Sigma F_y\right)^2} = \sqrt{\left(180 \text{ N}\right)^2 + \left(390 \text{ N}\right)^2} = 430 \text{ N}$$

$$\theta = \tan^{-1}\left(\frac{\Sigma F_y}{\Sigma F_x}\right) = \tan^{-1}\left(\frac{390 \text{ N}}{180 \text{ N}}\right) = 65.2°$$

or $\vec{F}_R = 430$ N at 65.2° north of east

Then, the resultant acceleration of the boat is

$$\vec{a}_R = \frac{\vec{F}_R}{m} = \frac{430 \text{ N}}{270 \text{ kg}} \text{ at } 65.2° \text{ north of east} = 1.59 \text{ m/s}^2 \text{ at } 65.2° \text{ north of east} \qquad \Diamond$$

Alternatively, we may first compute the components of the acceleration as

$$a_x = \frac{\Sigma F_x}{m} = \frac{180 \text{ N}}{270 \text{ kg}} = 0.667 \text{ m/s}^2 \text{ and } a_y = \frac{\Sigma F_y}{m} = \frac{390 \text{ N}}{270 \text{ kg}} = 1.44 \text{ m/s}^2$$

The resultant acceleration is then

$$a_R = \sqrt{\left(a_x\right)^2 + \left(a_y\right)^2} = \sqrt{\left(0.667 \text{ m/s}^2\right)^2 + \left(1.44 \text{ m/s}^2\right)^2} = 1.59 \text{ m/s}^2$$

$$\theta = \tan^{-1}\left(\frac{a_y}{a_x}\right) = \tan^{-1}\left(\frac{1.44 \text{ m/s}^2}{0.667 \text{ m/s}^2}\right) = 65.2°$$

or $\vec{a}_R = 1.59$ m/s² at 65.2° north of east $\Diamond$

23.   A 5.0-kg bucket of water is raised from a well by a rope. If the upward acceleration of the bucket is 3.0 m/s², find the force exerted by the rope on the bucket.

Solution

Choosing upward as positive, the acceleration of the bucket is

$$a_y = +3.0 \text{ m/s}^2$$

Newton's second law then gives the resultant vertical force acting on the bucket as

$$(F_R)_y = ma_y = (5.0 \text{ kg})(3.0 \text{ m/s}^2) = 15 \text{ N}$$

This resultant is also given by $(F_R)_y = \Sigma F_y = T - mg$

Thus, $T = (F_R)_y + mg = 15 \text{ N} + (5.0 \text{ kg})(9.80 \text{ m/s}^2) = 64 \text{ N}$ ◊

27. Assume that the three blocks portrayed in Figure P4.27 move on a frictionless surface and that a 42-N force acts as shown on the 3.0-kg block. Determine (a) the acceleration given this system, (b) the tension in the cord connecting the 3.0-kg and the 1.0-kg blocks, and (c) the force exerted by the 1.0-kg block on the 2.0-kg block.

Solution

(a) If we consider our system of interest to consist of all three blocks plus the connecting rope, the total mass of the system is

$$m_{total} = (1.0 + 2.0 + 3.0)\ kg = 6.0\ kg$$

The only horizontal external force acting on this system is the applied 42-N force directed toward the right (chosen as the positive x direction). Thus, Newton's second law gives

$$a_x = \frac{\Sigma F_x}{m_{total}} = \frac{42\ N}{6.0\ kg} = 7.0\ m/s^2\ \text{toward the right} \qquad \Diamond$$

(b) To determine the tension in the rope, define a new system consisting of only the 3.0-kg block. The external horizontal forces acting on this system are the applied 42-N force and the tension in the rope as shown in the free-body diagram at the right. Then,

$$\Sigma F_x = ma_x \implies 42\ N - T = m_3 a_x$$

or $T = 42\ N - m_3 a_x = 42\ N - (3.0\ kg)(7.0\ m/s^2) = 21\ N \qquad \Diamond$

(c) If we now consider a new system consisting of only the 2.0-kg block, the horizontal force exerted on it by the 1.0-kg block will be an external force as shown in the free-body diagram at the right. Applying Newton's second law to this system gives

$$\Sigma F_x = ma_x \implies F_{12} = m_2 a_x = (2.0\ kg)(7.0\ m/s^2) = 14\ N \qquad \Diamond$$

34. Two objects with masses of 3.00 kg and 5.00 kg are connected by a light string that passes over a frictionless pulley, as in Figure P4.34. Determine (a) the tension in the string, (b) the acceleration of each object, and (c) the distance each object will move in the first second of motion if both objects start from rest.

Solution

Examining the system shown in Figure P4.34, we realize that the 5.00-kg object will have a downward acceleration of some magnitude a. The 3.00-kg object will then accelerate upward at the same rate a. Consider a free-body diagram of each object:

For the 5.00-kg object:

$$\Sigma F_y = ma_y \Rightarrow T - 49.0\ \text{N} = (5.00\ \text{kg})(-a)$$

or $\qquad T = 49.0\ \text{N} - (5.00\ \text{kg})a$ $\qquad\qquad$ (1)

For the 3.00-kg object:

$$\Sigma F_y = ma_y \Rightarrow T - 29.4\ \text{N} = (3.00\ \text{kg})(+a)$$

or $\qquad T = 29.4\ \text{N} + (3.00\ \text{kg})a$ $\qquad\qquad$ (2)

Substituting Equation (2) into Equation (1) gives

$$29.4\ \text{N} + (3.00\ \text{kg})a = 49.0\ \text{N} - (5.00\ \text{kg})a$$

or $\qquad a = \dfrac{49.0\ \text{N} - 29.4\ \text{N}}{5.00\ \text{kg} + 3.00\ \text{kg}} = 2.45\ \text{m/s}^2$

(a) From Equation (1), the tension in the string is

$$T = 49.0\ \text{N} - (5.00\ \text{kg})(2.45\ \text{m/s}^2) = 36.8\ \text{N}$$ ◊

(b) The 5.00-kg object accelerates downward at $2.45\ \text{m/s}^2$, and the 3.00-kg object accelerates upward at $2.45\ \text{m/s}^2$ ◊

(c) If both objects start from rest $(v_{0y} = 0)$, the vertical distance each will move in the first second of motion given by $\Delta y = v_{0y}t + \frac{1}{2}a_y t^2$ as

$$|\Delta y| = 0 + \tfrac{1}{2}at^2 = \tfrac{1}{2}(2.45\ \text{m/s}^2)(1.00\ \text{s})^2 = 1.23\ \text{m}$$ ◊

37. A 1 000-N crate is being pushed across a level floor at a constant speed by a force $\vec{F}$ of 300 N at an angle of 20.0° below the horizontal, as shown in Figure P4.37a. (a) What is the coefficient of kinetic friction between the crate and the floor? (b) If the 300 N force is instead pulling the block at an angle of 20.0° above the horizontal, as shown in Figure 4.37b, what will be the acceleration of the crate. Assume that the coefficient of friction is the same as that found in (a).

Solution

(a) Figure (a) at the right is a free-body diagram of the crate in Figure P4.37a of the textbook. The crate is in equilibrium since it moves at constant velocity. Looking at the vertical forces, we can find the normal force:

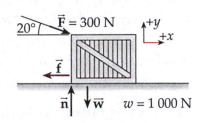

$$\Sigma F_y = +n - w - F\sin 20.0° = 0$$

$$n = 1\,000 \text{ N} + (300 \text{ N})\sin 20.0° = 1.10 \times 10^3 \text{ N}$$

Figure (a)

Then, considering the horizontal forces gives:

$$\Sigma F_x = +F\cos 20.0° - f = 0$$

$$f = (300 \text{ N})\cos 20.0° = 282 \text{ N}$$

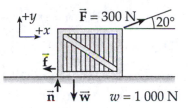

Therefore, the coefficient of kinetic friction between the crate and floor is given by:

Figure (b)

$$\mu_k = \frac{f}{n} = \frac{282 \text{ N}}{1.10 \times 10^3 \text{ N}} = 0.256$$

◊

(b) If the 300-N force pulls upward at 20.0° above the horizontal, the free-body diagram is as given in Figure (b). In this case, the vertical acceleration a_y is still zero, but the crate has some unknown horizontal acceleration. Considering the vertical forces,

$$\Sigma F_y = ma_y = 0 \qquad \text{giving} \qquad +F\sin 20.0° + n - w = 0$$

$$\text{or} \qquad n = 1\,000 \text{ N} - (300 \text{ N})\sin 20.0° = 897 \text{ N}$$

Then, assuming the same coefficient of friction as found in part (a), the friction force f is given by $f = \mu_k n = (0.256)(897 \text{ N}) = 230 \text{ N}$. Noting that the mass of the crate is $m = w/g = (1\,000 \text{ N})/(9.80 \text{ m/s}^2) = 102 \text{ kg}$, apply Newton's second law to the horizontal motion of the crate: $\Sigma F_x = ma_x$

$$\text{or } a_x = \frac{\Sigma F_x}{m} = \frac{F\cos 20.0° - f}{m} = \frac{(300 \text{ N})\cos 20.0° - (230 \text{ N})}{102 \text{ kg}} = 0.509 \text{ m/s}^2 \qquad ◊$$

45.   Objects with masses $m_1 = 10.0$ kg and $m_2 = 5.00$ kg are connected by a light string that passes over a frictionless pulley as in Figure P4.30. If, when the system starts from rest, m_2 falls 1.00 m in 1.20 s, determine the coefficient of kinetic friction between m_1 and the table.

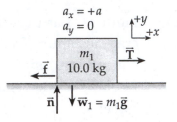

Figure P4.30

Solution

The free-body diagrams of the two objects in this system are shown at the right. Note that the accelerations of the two objects have the same magnitude, a, with the acceleration of m_1 directed horizontally to the right and the acceleration of m_2 directed vertically downward.

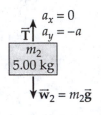

Since m_2 is observed to drop downward 1.00 m in 1.20 s when the system is released, the magnitude of the acceleration is found using $\Delta y = v_{0y}t + \frac{1}{2}a_y t^2$ as

$$-1.00 \text{ m} = 0 + \tfrac{1}{2}(-a)(1.20 \text{ s})^2 \text{ so } a = 1.39 \text{ m/s}^2$$

The weights of the objects are $w_1 = m_1 g = 98.0$ N and $w_2 = m_2 g = 49.0$ N. Applying Newton's second law to m_2 gives us the tension T in the cord:

$$\sum F_y = m_2 a_y: \qquad +T - w_2 = m_2(-a)$$

$$T = 49.0 \text{ N} + (5.00 \text{ kg})(-1.39 \text{ m/s}^2) = 42.1 \text{ N}$$

Now, consider the vertical forces acting on m_1:

$$\sum F_y = m_1 a_y: \qquad +n - w_1 = 0 \text{ so } n = w_1 = 98.0 \text{ N}$$

Finally, considering the horizontal forces acting on m_1,

$$\sum F_x = m_1 a_x: \qquad +T - f = m_1(+a)$$

$$f = T - m_1 a = 42.1 \text{ N} - (10.0 \text{ N})(1.39 \text{ m/s}^2) = 28.2 \text{ N}$$

The coefficient of kinetic friction between m_1 and the table is therefore

$$\mu_k = \frac{f}{n} = \frac{28.2 \text{ N}}{98.0 \text{ N}} = 0.287$$

51. The person in Figure P4.51 weighs 170 lb. Each crutch makes an angle of 22.0° with the vertical (as seen from the front). Half of the person's weight is supported by the crutches, the other half by the vertical forces exerted by the ground on his feet. Assuming that he is at rest and that the force exerted by the ground on the crutches acts along the crutches, determine (a) the smallest possible coefficient of friction between crutches and ground and (b) the magnitude of the compression force supported by each crutch.

Solution

The forces acting on the person's body are shown in the diagram at the right. $\vec{F}$ and $\vec{F'}$ are exerted by the two crutches, $\vec{n}_b$ is a normal force exerted by the ground as it directly supports half the person's weight ($n_b = w/2$), and $\vec{w}$ is the person's weight. The body is in equilibrium, so $a_x = a_y = 0$. Thus,

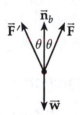

$$\Sigma F_x = 0 \implies F \sin\theta - F' \sin\theta = 0 \quad \text{or} \quad F' = F$$

Also $\qquad \Sigma F_y = 0 \implies F \cos\theta + F' \cos\theta + n_b - w = 0$

With $F' = F$ and $n_b = w/2$, this reduces to

$$2F \cos\theta + w/2 - w = 0 \quad \text{or} \quad F = \frac{w}{4\cos\theta}$$

Consider the free-body diagram of a crutch tip given at the right. Here $\vec{F}$ is the reaction force exerted on the crutch by the body and transmitted as a compression force to the tip. $\vec{n}$ and $\vec{f}_s$ are the normal force and static friction force exerted on the tip by the ground.

$$\Sigma F_y = 0 \implies n = F\cos\theta = \left(\frac{w}{4\cos\theta}\right)\cos\theta = \frac{w}{4}$$

$$\Sigma F_x = 0 \implies f_s = F\sin\theta = \left(\frac{w}{4\cos\theta}\right)\sin\theta = \frac{w}{4}\tan\theta$$

If the coefficient of friction between the crutch tip and the ground is the minimum necessary to prevent slipping, then $f_s = (f_s)_{max} = \mu_s n$ and

$$\mu_s = \frac{f_s}{n} = \frac{(w/4)\tan\theta}{w/4} = \tan\theta$$

(a)   With the crutch at angle $\theta = 22.0°$ from the vertical, the minimal coefficient of friction to prevent slipping is $\mu_s = \tan 22.0° = 0.404$ ◊

(b)   With $w = 170$ lbs and $\theta = 22.0°$, the compression force supported by the crutch is

$$F = \frac{w}{4\cos\theta} = \frac{170 \text{ lbs}}{4\cos(22.0°)} = 45.8 \text{ lbs} \qquad \Diamond$$

---

57.   A boy coasts down a hill on a sled, reaching a level surface at the bottom with a speed of 7.0 m/s. If the coefficient of friction between the sled's runners and the snow is 0.050 and the boy and sled together weigh 600 N, how far does the sled travel on the level surface before coming to rest?

Solution

The forces acting on the boy and sled are shown in the sketch. Applying Newton's second law gives:

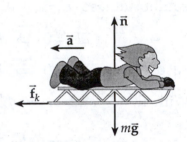

$$\Sigma F_y = ma_y \quad \Rightarrow \quad n - mg = 0 \quad \text{or} \quad n = mg$$

Thus, $f_k = \mu_k n = \mu_k mg$

$$\Sigma F_x = ma_x \quad \Rightarrow \quad -\mu_k mg = ma_x \quad \text{or} \quad a_x = -\mu_k g$$

Then, $v_x^2 = v_{0x}^2 + 2a_x \Delta x$ with $v_x = 0$ gives the distance the sled with initial speed v_{0x} will coast before stopping. This yields

$$\Delta x = \frac{0 - v_{0x}^2}{2a_x} = \frac{-v_{0x}^2}{2(-\mu_k g)} = \frac{v_{0x}^2}{2\mu_k g}$$

Therefore, if $v_{0x} = 7.0$ m/s and $\mu_k = 0.050$ the coasting distance on the level ground is

$$\Delta x = \frac{(7.0 \text{ m/s})^2}{2(0.050)(9.80 \text{ m/s}^2)} = 50 \text{ m} \qquad \Diamond$$

Note that the numeric value of the weight of the child and sled was never used in this solution. This means that any coaster starting with the given initial speed will go the same distance before stopping provided the coefficient of kinetic friction between runners and snow remains unchanged.

---

61. A frictionless plane is 10.0 m long and inclined at 35.0°. A sled starts at the bottom with an initial speed of 5.00 m/s up the incline. When the sled reaches the point at which it momentarily stops, a second sled is released from the top of the incline with an initial speed v_i. Both sleds reach the bottom of the incline at the same moment. (a) Determine the distance that the first sled traveled up the incline. (b) Determine the initial speed of the second sled.

Solution

First, consider the free-body diagram given at the right of either sled on the frictionless slope of inclination $\theta = 35.0°$. The acceleration of the sled will be directed down the incline (chosen as the +x direction) and has magnitude of

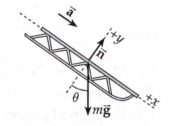

$$a_x = \frac{\Sigma F_x}{m} = \frac{mg \sin\theta}{m} = g \sin\theta$$

(a) If the first sled starts up the incline with speed $v_{0x} = -5.00 \, \text{m/s}$ at the bottom, the distance it travels up the incline before stopping is

$$|\Delta x| = \frac{\left| v_x^2 - v_{0x}^2 \right|}{2a_x} = \frac{\left| 0 - (-5.00 \, \text{m/s})^2 \right|}{2\left[(9.80 \, \text{m/s}^2) \sin 35.0° \right]} = 2.22 \, \text{m} \qquad \Diamond$$

(b) After stopping momentarily ($v_{0x} = 0$), the time for the first sled to go 2.22 m back down the incline is given by $\Delta x = v_{0x} t + \frac{1}{2} a_x t^2$ as

$$t = \sqrt{\frac{2\Delta x}{a_x}} = \sqrt{\frac{2(2.22 \, \text{m})}{(9.80 \, \text{m/s}^2) \sin 35.0°}} = 0.890 \, \text{s}$$

In this same time, the second sled (with $v_{0x} = +v_i$ and $a_x = g \sin\theta$) must travel the full length of the incline ($\Delta x = 10.0 \, \text{m}$). The required initial speed is given by $\Delta x = v_{0x} t + \frac{1}{2} a_x t^2$ as

$$v_i = \frac{\Delta x}{t} - \frac{1}{2} a_x t = \frac{10.0 \, \text{m/s}}{0.890 \, \text{s}} - \frac{1}{2} \left[(9.80 \, \text{m/s}^2) \sin 35.0° \right] (0.890 \, \text{s})$$

or $v_i = 8.74 \, \text{m/s}$ $\qquad \Diamond$

74.   On an airplane's takeoff, the combined action of the air around the engines and wings of an airplane exerts an 8 000-N force on the plane, directed upward at an angle of 65.0° above the horizontal. The plane rises with constant velocity in the vertical direction while continuing to accelerate in the horizontal direction. (a) What is the weight of the plane? (b) What is its horizontal acceleration?

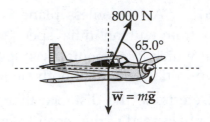

Solution

The weight of the plane, $\vec{w} = m\vec{g}$, is the only force acting on the plane other than the 8 000-N resultant force exerted by the engines and wings. Resolving the 8 000-N force into its horizontal and vertical components gives

$$F_x = (8\ 000\ \text{N})\cos 65.0° = 3.38 \times 10^3\ \text{N}$$

and $$F_y = (8\ 000\ \text{N})\sin 65.0° = 7.25 \times 10^3\ \text{N}$$

(a)   Since the plane has constant velocity in the vertical direction (that is, $a_y = 0$), $\Sigma F_y = m a_y$ yields

$$+F_y - w = 0 \qquad \text{or} \qquad w = F_y = 7.25 \times 10^3\ \text{N} \qquad\qquad \Diamond$$

(b)   The mass of the airplane is found as $m = \dfrac{w}{g} = \dfrac{7.25 \times 10^3\ \text{N}}{9.80\ \text{m}/\text{s}^2} = 740\ \text{kg}$

Then, applying Newton's second law to the horizontal motion gives:

$$a_x = \frac{\Sigma F_x}{m} = \frac{3.38 \times 10^3\ \text{N}}{740\ \text{kg}} = 4.57\ \text{m}/\text{s}^2 \qquad\qquad \Diamond$$

77. The board sandwiched between two other boards in Figure P4.77 weighs 95.5 N. If the coefficient of friction between the boards is 0.663, what must be the magnitude of the compression forces (assumed to be horizontal) acting on both sides of the center board to keep it from slipping?

$w = 95.5$ N

Figure P4.77

Solution

Since the board is in equilibrium, $\Sigma F_x = 0$ and we see that the normal forces must be the same on both sides of the board. Also, if the minimum normal forces (compression forces) are being applied, the board is on the verge of slipping and the friction force on each side is

$$f = (f_s)_{max} = \mu_s n$$

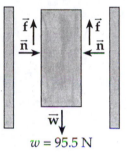

$w = 95.5$ N

The board is also in equilibrium in the vertical direction, so

$$\Sigma F_y = 2f - w = 0 \qquad \text{or} \qquad f = \frac{w}{2}$$

The minimum compression force needed is then

$$n = \frac{f}{\mu_s} = \frac{w}{2\mu_s} = \frac{95.5 \text{ N}}{2(0.663)} = 72.0 \text{ N}$$

◊

Chapter 5
Energy

NOTES FROM SELECTED CHAPTER SECTIONS

5.1 Work

In order for work to be accomplished, an object must undergo a displacement; the force associated with the work must have a component parallel to the direction of the displacement. Work is a scalar quantity and can be either positive or negative (positive when the component $F\cos\theta$ is in the same direction as the displacement). The SI unit of work is the newton-meter (N·m) or joule (J).

5.2 Kinetic Energy and the Work-Energy Theorem

Any object which has mass m and speed v has **kinetic energy**. Kinetic energy is a scalar quantity and has the same units as work; kinetic energy of an object will change only if net work is done on the object by external forces. The relationship between work and change in kinetic energy is stated in the work-energy theorem.

A force is **conservative** if the work it does on an object moving between two points is independent of the path the object takes between the points. The work done on an object by a conservative force depends only on the initial and final positions of the object. The gravitational force is an example of a conservative force.

A force is **non-conservative** if the work it does on an object moving between two points depends on the path taken. Kinetic friction is an example of a non-conservative force.

5.3 Gravitational Potential Energy

The work done on an object by the force of gravity is equal to the object's initial potential energy minus its final potential energy ($W_g = -\Delta PE_g$). The gravitational potential energy associated with an object depends only on the object's weight and its vertical height above the surface of the Earth. If the height above the surface increases, the potential energy will also increase; but the work done by the gravitational force will be negative. (In this case the direction of the displacement is opposite the direction of the gravitational force.) **In working problems involving gravitational potential energy, it is necessary to choose an arbitrary reference level (or location) at which the potential energy is taken to be zero.**

5.4   Spring Potential Energy

Elastic potential energy is the energy associated with a spring that is compressed or stretched beyond its equilibrium position. The spring force is a conservative force and has a magnitude that is proportional to the displacement of the spring from the equilibrium position.

5.5   Systems and Energy Conservation

The **change in the kinetic energy** of a physical system equals the sum of the work done by conservative forces and the work done by non-conservative forces. The sum of the kinetic energy plus the potential energy is called the **total mechanical energy**. Since the work done by conservative forces equals the negative of the change in potential energy ($W_c = -\Delta PE$), the work done by non-conservative forces equals the change in the total mechanical energy of the system. *The mechanical energy of a system remains constant if only conservative forces do work on the system.*

5.6   Power

Power delivered to an object is defined as the rate at which energy is transferred to the object or the rate at which work is being done on the object. The average power delivered to an object during a time interval can be expressed as the product of the average speed during the time interval and the component of the force in the direction of the velocity.

5.8   Work Done by a Varying Force

If the value of a non-constant force is known as a function of position, the work done by the varying force during a displacement can be determined by calculating the area under the force vs displacement curve.

EQUATIONS AND CONCEPTS

The **work done by a constant force** $\vec{F}$, (constant in both magnitude and direction) is defined to be the product of the component of the force in the direction of the displacement and the magnitude of the displacement.

$$W \equiv (F\cos\theta)\Delta x \qquad (5.2)$$

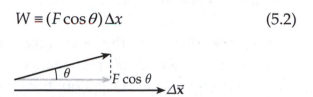

Work done by a variable force is equal to the area under the force vs displacement curve. The figure at right illustrates the case of a force acting along the x-axis.

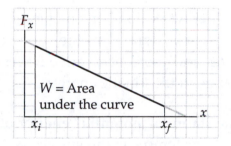

The work done by a force can be positive, negative, or zero, depending on the value of θ, the angle between the direction of the force and the direction of the displacement. The figure at the right shows an object acted on by four forces while being displaced down a smooth incline. The direction of the x-axis is taken to be along the direction of the incline. Consider the following cases:

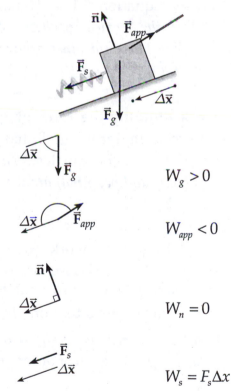

- If $0 \le \theta < 90°$, W is positive.

- If $90° < \theta \le 180°$, W is negative.

- If $\theta = 90°$ ($\vec{F}$ perpendicular to $\Delta\vec{x}$), then $W = 0$.

- If $\theta = 0$ ($\vec{F}$ and $\Delta\vec{x}$ are parallel and in the same direction), $W = F\Delta x$.

$$W_g > 0$$

$$W_{app} < 0$$

$$W_n = 0$$

$$W_s = F_s\Delta x$$

Work is a scalar quantity and the SI unit of work is the newton-meter or joule.

$1\,\text{N}\cdot\text{m} = 1\,\text{J}$

joule (J) = newton $\cdot$ meter = $\text{kg}\cdot\text{m}^2/\text{s}^2$

Kinetic energy, defined by Eq. 5.5, is a scalar quantity and is the energy associated with the motion of a mass m.

$$KE \equiv \tfrac{1}{2}mv^2 \tag{5.5}$$

The **work done by the net force** can be expressed in terms of the change in the kinetic energy of the body. Equations 5.4 and 5.6 are statements of **the work-energy theorem**.

$$W_{net} = \tfrac{1}{2}mv^2 - \tfrac{1}{2}mv_0^2 \tag{5.4}$$

$$W_{net} = KE_f - KE_i = \Delta KE \tag{5.6}$$

The **net work** done on an object is the sum of the work done by the non-conservative forces and the conservative forces equals the change in the kinetic energy.

$$W_{nc} + W_c = \Delta KE \tag{5.7}$$

Gravitational potential energy of the mass-Earth system near the surface of the Earth is defined by Equation 5.10. *The quantity y in Eq. 5.10 is the vertical position of the mass relative to the Earth (or other arbitrarily chosen reference level where $PE_g = 0$).*

$$PE \equiv mgy \tag{5.10}$$

The **work done by the force of gravity** can be expressed in terms of initial and final values of the y-coordinates. *The units of energy (kinetic and potential) are the same as the units of work.*

$$W_g = -\left(PE_f - PE_i\right) \tag{5.11}$$

$$= -\left(mgy_f - mgy_i\right)$$

In calculating the work done by the gravitational force, choose an origin from which it is convenient to calculate PE_i and PE_f for a particular situation. *The difference in potential energy between two points is independent of the location of the origin.*

Reference level for gravitational potential energy

When only conservative forces act on a system, the total mechanical energy ($KE+PE$) of the system remains constant. *This is a statement of the law of conservation of mechanical energy.*

$$KE_i + PE_i = KE_f + PE_f \qquad (5.13)$$

The **restoring force** exerted by a stretched or compressed spring is proportional to the displacement from the equilibrium position of the "free" end of the spring and is directed opposite the displacement. *The factor k is called the spring constant or elastic constant and is characteristic of a particular spring.* The units of k are N/m.

$$F_s \equiv -kx \qquad (5.15)$$

Elastic potential energy represents the work done by a force in stretching or compressing a spring. For a given displacement from the equilibrium position, the potential energy in the spring depends on the spring constant, k.

$$PE_s = \tfrac{1}{2}kx^2 \qquad (5.16)$$

k = spring constant in N/m

The **work done by all non-conservative forces** equals the change in the total mechanical energy of a system (change in kinetic energy plus change in the potential energy).

$$W_{nc} + W_c = \Delta KE$$

$$W_{nc} = (KE_f + PE_f) - (KE_i + PE_i) \qquad (5.12)$$

The **average power** supplied by a force is the ratio of the work done by the force to the time interval over which the force acts. The average power can also be expressed in terms of the force and the average speed of the object on which the force acts. *In Equation 5.23, F is the component of the force along the direction of the velocity.*

$$\mathscr{P}_{av} \equiv \frac{W}{\Delta t} \qquad (5.22)$$

$$\mathscr{P}_{av} = F v_{av} \qquad (5.23)$$

The **SI unit of power** is the watt; in the U.S. customary system, the unit of power is the horsepower.

$$1\,W = 1\,J/s = 1\,kg \cdot m^2/s^3 \qquad (5.24)$$

$$1\,hp = 550\,\frac{ft \cdot lb}{s} = 746\,W \qquad (5.25)$$

SUGGESTIONS, SKILLS, AND STRATEGIES

Choosing a Zero Level for Potential Energy

In working problems involving gravitational potential energy, it is always necessary to choose a location at which to set the gravitational potential energy equal to zero. This choice is completely arbitrary because the important quantity is the difference in potential energy, and that difference is independent of the location of zero. It is often convenient, but not essential, to choose the surface of the Earth as the reference position for zero potential energy. In most cases, the statement of the problem suggests a convenient level to use.

Conservation of Energy

Take the following steps in applying the principle of conservation of energy:

1. Define your system, which may consist of more than one object.

2. Select a reference level for the zero point of gravitational potential energy. This level must not be changed during the solution of a specific problem.

3. Determine whether or not non-conservative forces are present.

4. If mechanical energy is conserved (that is, if only conservative forces are present), you should use equation 5.13 ($KE_i + PE_i = KE_f + PE_f$). Usually (except when a quantity is equal to zero) it is better to solve for the unknown algebraically (using symbols) before substituting numerical values.

5. If non-conservative forces such as friction are present (and thus mechanical energy is not conserved), first write expressions for the total initial and total final energies. In this case, the difference between the two total energies is equal to the work done by the non-conservative force(s). That is, you should apply Equation 5.12.

REVIEW CHECKLIST

▷ Define the work done by a constant force and work done by a force which varies with position. (Recognize that the work done by a force can be positive, negative, or zero; describe at least one example of each case).

▷ Understand that the work done by a conservative force in moving a body between any two points is independent of the path taken. Non-conservative forces are those for which the work done on a particle moving between two points depends on the path. Account for non-conservative forces acting on a system using the work-energy theorem. In this case, the work done by all non-conservative forces equals the change in total mechanical energy of the system.

▷ Relate the work done by the net force on an object to the **change** in kinetic energy. The relation $W_{net} = \Delta KE = KE_f - KE_i$ is called the work-energy theorem, and is valid whether or not the (resultant) force is constant. That is, if we know the net work done on a particle as it undergoes a displacement, we also know the change in its kinetic energy. This is the most important concept in this chapter, so you must understand it thoroughly.

▷ Recognize that the gravitational potential energy of the mass-Earth system, $PE_g = mgy$, can be positive, negative, or zero, depending on the location of the reference level used to measure y. Be aware of the fact that although PE depends on the origin of the coordinate system, the change in potential energy, $(PE)_f - (PE)_i$, is independent of the coordinate system used to define PE.

▷ Calculate average power when work is accomplished over a time interval and instantaneous power when an applied force acts on an object moving with speed v.

SOLUTIONS TO SELECTED END-OF-CHAPTER PROBLEMS

7.   A sledge loaded with bricks has a total mass of 18.0 kg and is pulled at constant speed by a rope inclined at 20.0° above the horizontal. The sledge moves a distance of 20.0 m on a horizontal surface. The coefficient of kinetic friction between the sledge and surface is 0.500. (a) What is the tension in the rope? (b) How much work is done by the rope on the sledge? (c) What is the mechanical energy lost due to friction?

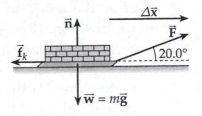

Solution

(a)   The sledge has constant velocity, so

$$\Sigma F_y = n + F \sin 20.0° - mg = ma_y = 0 \;\Rightarrow\; n = mg - F \sin 20.0°$$

Thus, the kinetic friction force is $f_k = \mu_k n = \mu_k \left(mg - F \sin 20.0° \right)$

Also, $\Sigma F_x = F \cos 20.0° - f_k = ma_x = 0$

or $F \cos 20.0° - \mu_k \left(mg - F \sin 20.0° \right) = 0$

Simplifying this result gives the tension in the rope as

$$F = \frac{\mu_k mg}{\cos 20.0° + \mu_k \sin 20.0°} = \frac{0.500 \left(18.0\ \text{kg}\right)\left(9.80\ \text{m/s}^2\right)}{\cos 20.0° + 0.500 \sin 20.0°} = 79.4\ \text{N}$$ ◊

(b)   The work done by the rope on the sledge is $W_F = \left(F \cos 20.0° \right) \Delta x$, or

$$W_F = \left(79.4\ \text{N}\right)\left(20.0\ \text{m}\right)\cos 20.0° = 1.49 \times 10^3\ \text{J} = 1.49\ \text{kJ}$$ ◊

(c)   The friction force and the displacement are in opposite directions ($\theta = 180°$), so the work done by friction on the sledge is

$$W_{f_k} = \left(f_k \cos 180° \right) \Delta x = \mu_k \left(mg - F \sin 20.0° \right)\left(\Delta x \right) \cos 180°$$

$$= 0.500 \left[\left(18.0\ \text{kg}\right)\left(9.80\ \text{m/s}^2\right) - \left(79.4\ \text{N}\right)\sin 20.0° \right]\left(20.0\ \text{m}\right)\left(-1\right)$$

or $W_{f_k} = -1.49 \times 10^3\ \text{J} = -1.49\ \text{kJ}$ ◊

13.   A 70-kg base runner begins his slide into second base when he is moving at a speed of 4.0 m/s. The coefficient of friction between his clothes and Earth is 0.70. He slides so that his speed is zero just as he reaches the base. (a) How much mechanical energy is lost due to friction acting on the runner? (b) How far does he slide?

Solution

(a)   Since the base runner slides on a level surface, the change in gravitational potential energy is

$$\Delta PE_g = mgy_f - mgy_i = mg(\Delta y) = 0$$

The change in mechanical energy is therefore

$$W_{nc} = (KE + PE)_f - (KE + PE)_i = (\Delta KE) + (\Delta PE) = \left(\tfrac{1}{2}mv_f^2 - \tfrac{1}{2}mv_i^2\right) + (0)$$

or $W_{nc} = \tfrac{1}{2}m\left(v_f^2 - v_i^2\right) = \dfrac{1}{2}(70 \text{ kg})\left[0 - (4.0 \text{ m/s})^2\right] = -5.6 \times 10^2 \text{ J}$ ◊

(b)   The friction force is directed opposite to the displacement ($\theta = 180°$) and has magnitude $f_k = \mu_k n = \mu_k mg$. All other forces acting on the base runner are perpendicular to the displacement and, hence, do no work. The total work done by non-conservative forces is then $W_{nc} = (f_k \cos 180°)\Delta x$, so the displacement is

$$\Delta x = \frac{W_{nc}}{f_k \cos 180°} = \frac{W_{nc}}{-\mu_k mg} = \frac{-5.6 \times 10^2 \text{ J}}{-(0.70)(70 \text{ kg})(9.80 \text{ m/s}^2)} = 1.2 \text{ m}$$ ◊

17.   A 2 000-kg car moves down a level highway under the actions of two forces: a 1 000-N forward force exerted on the drive wheels by the road and a 950-N resistive force. Use the work-energy theorem to find the speed of the car after it has moved a distance of 20 m, assuming that it starts from rest.

Solution

The forward force exerted on the car by the road is in the direction of the displacement $(\theta_1 = 0°)$ while the resistive force is directed opposite to the displacement $(\theta_2 = 180°)$.

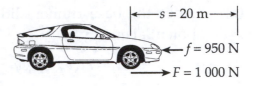

$s = 20$ m
$f = 950$ N
$F = 1\,000$ N

Other forces acting on the car (that is, the normal force exerted by the road and the gravitational force) are perpendicular to the displacement and, hence, do no work.

Therefore, the net work done on the car is $W_{net} = (F\cos\theta_1)s + (f\cos\theta_2)s$:

$$W_{net} = \left[(1\,000 \text{ N})\cos 0°\right]s + \left[(950 \text{ N})\cos 180°\right]s$$

$$W_{net} = (1\,000 \text{ N} - 950 \text{ N})(20 \text{ m}) = 1.0\times10^3 \text{ J}$$

The work-energy theorem then gives

$$W_{net} = \tfrac{1}{2}mv_f{}^2 - \tfrac{1}{2}mv_i{}^2 = \tfrac{1}{2}mv_f{}^2 - 0$$

or $$v_f = \sqrt{\frac{2W_{net}}{m}} = \sqrt{\frac{2(1.0\times10^3 \text{ J})}{2\,000 \text{ kg}}} = 1.0 \text{ m/s}$$

◊

22. Truck suspensions often have "helper springs" that engage at high loads. One such arrangement is a leaf spring with a helper coil spring mounted on the axle, as shown in Figure P5.22. When the main leaf spring is compressed by distance y_0, the helper spring engages and then helps to support any additional load. Suppose the leaf spring constant is 5.25×10^5 N/m, the helper spring constant is 3.60×10^5 N/m, and $y_0 = 0.500$ m. (a) What is the compression of the leaf spring for a load of 5.00×10^5 N? (b) How much work is done in compressing the springs?

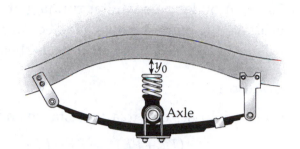

Figure P5.22

Solution

(a) When the load comes to equilibrium, the total upward support force exerted by the springs must equal the weight of the load, or

$$\left(F_s\right)_{total} = \left(F_s\right)_{leaf} + \left(F_s\right)_{helper} = w_{load}$$

If k_ℓ and k_h are the spring constants of the leaf spring and the helper spring, respectively, and x_ℓ is the vertical distance the ends of the leaf spring are displaced while $x_k = x_\ell - y_0$ is the vertical distance the helper spring is compressed, the condition for equilibrium becomes

$$k_\ell x_\ell + k_h \left(x_\ell - y_0\right) = w_{load}$$

The compression of the leaf spring at equilibrium is then

$$x_\ell = \frac{w_{load} + k_h y_0}{k_\ell + k_h} = \frac{5.00 \times 10^5 \text{ N} + \left(3.60 \times 10^5 \text{ N/m}\right)\left(0.500 \text{ m}\right)}{5.25 \times 10^5 \text{ N/m} + 3.60 \times 10^5 \text{ N/m}} = 0.768 \text{ m} \qquad \Diamond$$

(b) The work done compressing the springs is equal to the elastic potential energy stored in the springs at equilibrium, or

$$W_s = \tfrac{1}{2} k_\ell x_\ell^2 + \tfrac{1}{2} k_h \left(x_\ell - y_0\right)^2 = \tfrac{1}{2} k_\ell x_\ell^2 + \tfrac{1}{2} k_h \left(x_\ell - 0.500 \text{ m}\right)^2$$

$$W_s = \tfrac{1}{2}\left(5.25 \times 10^5 \text{ N/m}\right)\left(0.768 \text{ m}\right)^2 + \tfrac{1}{2}\left(3.60 \times 10^5 \text{ N/m}\right)\left(0.268 \text{ m}\right)^2 = 1.68 \times 10^5 \text{ J} \qquad \Diamond$$

26.   A 50-kg pole vaulter running at 10 m/s vaults over the bar. Her speed when she is above the bar is 1.0 m/s. Neglect air resistance, as well as any energy absorbed by the pole, and determine her altitude as she crosses the bar.

Solution

If energy losses due to air resistance and the flexing of the pole are negligible, then

$$(KE + PE)_f = (KE + PE)_i$$

or the total mechanical energy is conserved.

Consider the time interval from when the runner is approaching the bar at constant velocity until the instant she is directly above the bar. Since the pole is unbent as she runs toward the bar and is also unbent as she passes above the bar, there is no change in elastic potential energy of the system. Thus, the only change in potential energy occurs in gravitational potential energy, and conservation of mechanical energy gives

$$\tfrac{1}{2}\not{m}v_f^2 + \not{m}gy_f = \tfrac{1}{2}\not{m}v_i^2 + \not{m}gy_i$$

Her change in altitude as she goes over the bar is then

$$\Delta y = y_f - y_i = \frac{v_i^2 - v_f^2}{2g} = \frac{(10 \ \text{m/s})^2 - (1.0 \ \text{m/s})^2}{2(9.80 \ \text{m/s}^2)} = 5.1 \ \text{m} \qquad \Diamond$$

33. The launching mechanism of a toy gun consists of a spring of unknown spring constant, as shown in Figure P5.33a. If the spring is compressed a distance of 0.120 m and the gun fired vertically as shown, the gun can launch a 20.0-g projectile from rest to a maximum height of 20.0 m above the starting point of the projectile. Neglecting all resistive forces, determine (a) the spring constant and (b) the speed of the projectile as it moves through the equilibrium position of the spring (where $x = 0$), as shown in Figure P5.33b.

Solution

Since resistance forces may be neglected, the mechanical energy of the system (gun and ball) will be conserved:

$$\left(KE + PE_g + PE_s\right)_f = \left(KE + PE_g + PE_s\right)_i$$

(a) Choose the initial state to be the instant the ball starts from rest (with spring compressed) and the final state to be the instant the ball reaches maximum height (spring relaxed). With these choices, $v_i = v_f = 0$, $y_f - y_i = 20.0$ m, $x_i = 0.120$ m, and $x_f = 0$. The energy conservation equation then reduces to

$$0 + mgy_f + 0 = 0 + mgy_i + \tfrac{1}{2}kx_i^2$$

and gives the spring constant as

$$k = \frac{2mg\left(y_f - y_i\right)}{x_i^2} = \frac{2\left(20.0\times10^{-3}\text{ kg}\right)\left(9.80\text{ m/s}^2\right)\left(20.0\text{ m}\right)}{\left(0.120\text{ m}\right)^2} = 544\text{ N/m} \qquad \Diamond$$

(b) In this case, select the initial state to be as before but take the final state to be the instant the ball passes the $x = 0$ level. Then $y_f - y_i = 0.120$ m and energy conservation yields

$$\tfrac{1}{2}mv_f^2 + mgy_f + 0 = 0 + mgy_i + \tfrac{1}{2}kx_i^2 \qquad \text{or} \qquad v_f = \sqrt{\frac{kx_i^2}{m} - 2g\left(y_f - y_i\right)}$$

$$v_f = \sqrt{\frac{\left(544\text{ N/m}\right)\left(0.120\text{ m}\right)^2}{20.0\times10^{-3}\text{ kg}} - 2\left(9.80\text{ m/s}^2\right)\left(0.120\text{ m}\right)} = 19.7\text{ m/s} \qquad \Diamond$$

45.   A skier starts from rest at the top of a hill that is inclined 10.5° with respect to the horizontal. The hillside is 200 m long, and the coefficient of friction between snow and skis is 0.075 0. At the bottom of the hill, the snow is level and the coefficient of friction is unchanged. How far does the skier move along the horizontal portion of the snow before coming to rest?

Solution

Select the reference level of gravitational potential energy $\left(PE_g = 0\right)$ at the level of the base of the hill and let x represent the horizontal distance the skier travels after reaching this level.

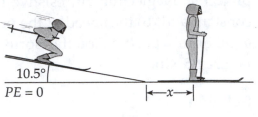

While on the hill, the normal force exerted on the skier by the snow is $n_1 = mg\cos 10.5°$ and the friction force is

$$f_1 = \mu_k n_1 = \mu_k mg\cos 10.5°$$

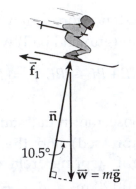

On the level snow, the normal force is $n_2 = mg$ and the friction force is $f_2 = \mu_k n_2 = \mu_k mg$

Consider the entire trip, from when the skier starts from rest on the hill until the skier comes to rest on the level snow. Then, $y_i - y_f = (200 \text{ m})\sin 10.5°$ and $v_i = v_f = 0$.

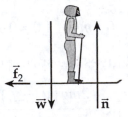

The work done by non-conservative forces is

$$W_{nc} = f_1 (200 \text{ m})\cos 180° + \left(f_2\right)(x)\cos 180° = -\mu_k mg\left[(200 \text{ m})\cos 10.5° + x\right]$$

Application of the work-energy theorem,

$$W_{nc} = \left(KE + PE_g\right)_f - \left(KE + PE_g\right)_i$$

then gives:

$$-\mu_k mg\left[(200 \text{ m})\cos 10.5° + x\right] = 0 + mgy_f - 0 - mgy_i = -mg(200 \text{ m})\sin 10.5°$$

or $x = (200 \text{ m})\left[\dfrac{\sin 10.5°}{\mu_k} - \cos 10.5°\right]$

With $\mu_k = 0.075\,0$, this yields $x = 289 \text{ m}$.   ◊

51.   The electric motor of a model train accelerates the train from rest to 0.620 m/s in 21.0 ms. The total mass of the train is 875 g. Find the average power delivered to the train during its acceleration.

Solution

The total work input to the train is equal to the change in the mechanical energy of the train, or

$$W_{input} = (KE + PE)_f - (KE + PE)_i$$

Assuming a level track, the potential energy of the train will remain constant $\left(PE_f = PE_i\right)$, and the above expression reduces to

$$W_{input} = KE_f - KE_i = \tfrac{1}{2}m\left(v_f^2 - v_i^2\right)$$

where m is the mass of the moving train. The average power delivered to the train as it is accelerated from $v_i = 0$ to $v_f = 0.620$ m/s in an elapsed time of $\Delta t = 21.0$ ms is

$$\mathcal{P}_{av} = \frac{W_{input}}{\Delta t} = \frac{\tfrac{1}{2}m\left(v_f^2 - v_i^2\right)}{\Delta t} = \frac{(0.875 \text{ kg})\left[(0.620 \text{ m/s})^2 - 0\right]}{2\left(21.0 \times 10^{-3} \text{ s}\right)} = 8.01 \text{ J/s} = 8.01 \text{ W} \qquad \Diamond$$

55.   The force acting on a particle varies as in Figure P5.55. Find the work done by the force as the particle moves (a) from $x = 0$ m to $x = 8.00$ m, (b) from $x = 8.00$ m to $x = 10.0$ m, and (c) from $x = 0$ m to $x = 10.0$ m.

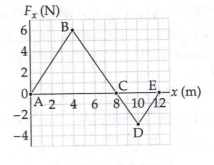

Solution

The work done on the particle by the force F as the particle moves from $x = x_i$ to $x = x_f$ is the area under the curve from x_i to x_f.

Figure P5.55

(a)   For $x = 0$ to $x = 8.00$ m

W = area of triangle $ABC = \frac{1}{2}\overline{AC} \times$ altitude

$W_{0 \to 8} = \frac{1}{2}(8.00 \text{ m})(6.00 \text{ N}) = 24.0$ J   ◊

(b)   For $x = 8.00$ m   to $x = 10.0$ m

$W_{8 \to 10}$ = area of triangle $CDE = \frac{1}{2}\overline{CE} \times$ altitude

$W_{8 \to 10} = \frac{1}{2}(2.00 \text{ m})(-3.00 \text{ N}) = -3.00$ J   ◊

(c)   For $x = 0$ to $x = 10.0$ m

$W_{0 \to 10} = W_{0 \to 8} + W_{8 \to 10} = 24.0 \text{ J} + (-3.00 \text{ J}) = 21.0$ J   ◊

62. A toy gun uses a spring to project a 5.3-g soft rubber sphere horizontally. The spring constant is 8.0 N/m, the barrel of the gun is 15 cm long, and a constant frictional force of 0.032 N exists between barrel and projectile. With what speed does the projectile leave the barrel if the spring was compressed 5.0 cm for this launch?

Solution

As the rubber sphere travels the length of the barrel of the gun, the only non-conservative force doing work on it is the frictional force. The normal force exerted by the barrel on the sphere is non-conservative, but is perpendicular to the displacement and does no work. The gravitational force and the spring force are both conservative, and their effects can be accounted for with potential energy terms. We write the work-energy theorem as

$$W_{nc} = \left(KE + PE_g + PE_s\right)_f - \left(KE + PE_g + PE_s\right)_i = \Delta KE + \Delta PE_g + \Delta PE_s$$

Since the gun barrel is horizontal, the gravitational potential energy is constant, or $\Delta PE_g = mg\left(y_f - y_i\right) = 0$. The elastic potential energy is $PE_s = \frac{1}{2}kx^2$ where x is the distance the spring is compressed, so $\Delta PE_s = \frac{1}{2}k\left(0 - x_i^2\right) = -\frac{1}{2}kx_i^2$.

The frictional force is always directed opposite to the displacement $(\theta = 180°)$, and the work it does is $W_{nc} = \left(f \cos 180°\right)L = -fL$ where L is the length of the gun barrel.

The work-energy theorem then becomes:

$$-fL = \left(\tfrac{1}{2}mv_f^2 - 0\right) + (0) + \left(-\tfrac{1}{2}kx_i^2\right) \quad \text{or} \quad v_f = \sqrt{\frac{kx_i^2 - 2fL}{m}}$$

and the speed of the rubber sphere as it leaves the barrel is

$$v_f = \sqrt{\frac{(8.0 \text{ N/m})(0.050 \text{ m})^2 - 2(0.032 \text{ N})(0.15 \text{ m})}{5.3 \times 10^{-3} \text{ kg}}} = 1.4 \text{ m/s} \qquad \Diamond$$

75.   A child's pogo stick (Fig. P5.75) stores energy in a spring ($k = 2.50 \times 10^4$ N/m). At position A ($x_1 = -0.100$ m), the spring compression is a maximum and the child is momentarily at rest. At position B ($x = 0$), the spring is relaxed and the child is moving upwards. At position C, the child is again momentarily at rest at the top of the jump. Assuming that the combined mass of child and pogo stick is 25.0 kg, (a) calculate the total energy of the system if both potential energies are zero at $x = 0$, (b) determine x_2, (c) calculate the speed of the child at $x = 0$, (d) determine the value of x for which the kinetic energy of the system is a maximum, and (e) obtain the child's maximum upward speed.

Figure P5.75

Solution

We choose $PE_g = 0$ at the level where the spring is relaxed ($x = 0$), or at the level of position B.

(a)   At position A, $v = 0$ and the total energy of the system is given by

$$E = \left(KE + PE_g + PE_s\right)_A = \frac{1}{2}mv_A^2 + mg\,x_1 + \frac{1}{2}k\,x_1^2, \text{ or}$$

$$E = 0 + (25.0 \text{ kg})(9.80 \text{ m/s}^2)(-0.100 \text{ m}) + \frac{1}{2}(2.50 \times 10^4 \text{ N/m})(-0.100 \text{ m})^2$$

$$E = 101 \text{ J} \qquad\qquad\qquad \Diamond$$

(b)   In position C, $v = 0$ and the spring is uncompressed giving $KE = 0$ and $PE_s = 0$

Hence, $$E = \left(0 + PE_g + 0\right)_C = mg\,x_2$$

or $$x_2 = \frac{E}{mg} = \frac{101 \text{ J}}{(25.0 \text{ kg})(9.80 \text{ m/s}^2)} = 0.410 \text{ m} \qquad\qquad \Diamond$$

(c)   At Position B, $PE_g = PE_s = 0$ and $E = (KE + 0 + 0)_B = \frac{1}{2}mv_B^2$

Therefore, $$v_B = \sqrt{\frac{2E}{m}} = \sqrt{\frac{2(101 \text{ J})}{25.0 \text{ kg}}} = 2.84 \text{ m/s} \qquad\qquad \Diamond$$

(d)   Where the velocity (and hence kinetic energy) is a maximum, the slope of the velocity versus time graph (that is, the acceleration) must be zero. Thus, $\Sigma F_y = ma_y = 0$ or the upward force due to the spring must exactly counterbalance the downward gravitational force where the kinetic energy is a maximum. Taking upward as positive, the position of interest is defined by $\Sigma F_y = -kx - mg = 0$, or

$$x = -\frac{mg}{k} = -\frac{245 \text{ kg}}{2.50 \times 10^4 \text{ N/m}} = -9.80 \times 10^{-3} \text{ m} = -9.80 \text{ mm}$$ ◊

(e)   From the total energy, $E = KE + PE_g + PE_s = \frac{1}{2}mv^2 + mg\,x + \frac{1}{2}k\,x^2$, we find

$$v = \sqrt{\frac{2E}{m} - 2g\,x - \frac{k}{m}x^2}$$

Where the speed, and hence kinetic energy is a maximum (that is, at $x = -9.80 \times 10^{-3}$ m), this gives

$$v_{max} = 2.85 \text{ m/s}$$ ◊

81.   A loaded ore car has a mass of 950 kg and rolls on rails with negligible friction. It starts from rest and is pulled up a mine shaft by a cable connected to a winch. The shaft is inclined at 30.0° above the horizontal. The car accelerates uniformly to a speed of 2.20 m/s in 12.0 s and then continues at constant speed. (a) What power must the winch motor provide when the car is moving at constant speed? (b) What maximum power must the motor provide? (c) What total energy transfers out of the motor by work by the time the car moves off the end of the track, which is of length 1 250 m?

Solution

(a)   When the car is being towed up the incline at constant speed, the tension in the cable is $F = mg \sin\theta$, and the power input from the motor is

$$\mathcal{P} = Fv = mgv \sin\theta$$

or $\mathcal{P} = (950 \text{ kg})(9.80 \text{ m/s}^2)(2.20 \text{ m/s})\sin 30.0° = 1.02 \times 10^4 \text{ W} = 10.2 \text{ kW}$ ◊

(b)   While the car is accelerating, the tension in the cable is

$$F_a = mg \sin\theta + ma = m\left(g\sin\theta + \frac{\Delta v}{\Delta t}\right)$$

$$= (950 \text{ kg})\left[(9.80 \text{ m/s}^2)\sin 30.0° + \frac{2.20 \text{ m/s} - 0}{12.0 \text{ s}}\right] = 4.83 \times 10^3 \text{ N}$$

The power input, $\mathcal{P} = Fv$, is a maximum at the instant when both the tension in the cable and the speed of the car have their maximum values. The maximum tension $(F = F_a)$ occurs as the car is being accelerated, and maximum speed $(v_{max} = 2.20 \text{ m/s})$ also occurs at the last instant of the acceleration phase. Thus, the maximum power from the motor is

$$\mathcal{P}_{max} = F_a v_{max} = (4.83 \times 10^3 \text{ N})(2.20 \text{ m/s}) = 10.6 \text{ kW}$$ ◊

(c)   The work the motor does moving the car up the frictionless track is

$$W_{nc} = (KE + PE)_f - (KE + PE)_i = \tfrac{1}{2}mv_f^2 + mg(y_f - y_i) = m\left[\tfrac{1}{2}v_f^2 + g(L\sin\theta)\right]$$

or $W_{nc} = (950 \text{ kg})\left[\tfrac{1}{2}(2.20 \text{ m/s})^2 + (9.80 \text{ m/s}^2)(1\,250 \text{ m})\sin 30.0°\right] = 5.82 \times 10^6 \text{ J}$ ◊

Chapter 6
Momentum and Collisions

NOTES FROM SELECTED CHAPTER SECTIONS

6.1 Momentum and Impulse

The **time rate of change of the momentum** of a particle is equal to the resultant force on the particle. The **impulse** of a force is a vector quantity and is equal to the change in momentum of the particle on which the force acts. The impulse or change in momentum of an object is equal to the area under the force vs time graph from the beginning to the end of the time interval during which the force is in contact with the object. Under the **impulse approximation,** it is assumed that one of the forces acting on a particle is of short time duration but of much greater magnitude than any of the other forces.

6.2 Conservation of Momentum

6.3 Collisions

The principle of **conservation of linear momentum** can be stated for an isolated system of objects. This is a system on which no external forces (e.g. friction or gravity) are acting. **When no external forces act on a system the total linear momentum of the system remains constant.** Remember, momentum is a vector quantity and the momentum of each individual particle may change but the total momentum of the entire system of particles remains constant. A collision between two or more masses is an important example of conservation of momentum.

For **any type of collision**, the total momentum before the collision equals the total momentum just after the collision.

In an **inelastic collision,** the total momentum is conserved; however, the total kinetic energy is not conserved.

In a **perfectly inelastic** collision, the two colliding objects stick together following the collision. This corresponds to a maximum loss in kinetic energy.

In an **elastic collision,** both momentum and kinetic energy are conserved.

6.4   Glancing Collisions

The law of conservation of momentum is not restricted to one-dimensional collisions. If two masses undergo a **two-dimensional** (glancing) **collision** and there are no external forces acting, the total momentum is conserved in both the *x and y* directions independently.

EQUATIONS AND CONCEPTS

An object of mass *m* and velocity $\vec{v}$ is characterized by a vector quantity called **linear momentum**. The SI units of linear momentum are kg·m/s. Since momentum is a vector quantity, the defining equation can be written in component form.

$$\vec{p} = m\vec{v} \qquad (6.1)$$

$$p_x = mv_x$$

$$p_y = mv_y$$

The **impulse** imparted to an object is the product of the net force acting on the object and the time interval during which it acts. *Impulse is a vector quantity and has the same direction as the applied force. The SI unit of impulse is kilogram meter per second.*

$$\vec{I} \equiv \vec{F}\Delta t \qquad (6.4)$$

The impulse imparted by a force during a time interval Δt is equal to the area under the force vs time graph from the beginning to the end of the time interval.

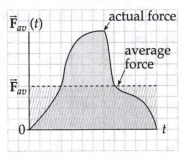

When the force varies in time, as illustrated in the figure, it is often convenient to define an **average force,** which is a constant force that imparts the same impulse in a time interval Δt as the actual time varying force.

$$\vec{F}_{av}\Delta t = \Delta\vec{p} \qquad (6.6)$$

The **principle of conservation of momentum** is stated in Equation 6.7. *When two objects interact in a collision (and exert forces on each other) and no external forces act on the two-object system, the total momentum of the system before the collision equals the total momentum after the collision*

$$m_1\vec{v}_{1i} + m_2\vec{v}_{2i} = m_1\vec{v}_{1f} + m_2\vec{v}_{2f} \qquad (6.7)$$

Conservation of momentum for a two-body collision

In general, when the resultant external force acting on any system containing any number of interacting objects is zero, the linear momentum of the system is conserved.

Law of conservation of linear momentum

An **elastic collision** is one in which both momentum and kinetic energy are conserved.

Types of collisions

An **inelastic collision** is one in which momentum is conserved but kinetic energy is not.

A **perfectly inelastic collision** is a collision in which the colliding objects stick together so that they have a common final velocity and the momentum of the system of objects is conserved.

Conservation of momentum for a one-dimensional perfectly inelastic collision. Note that v_{1i}, v_{2i}, and v_f are actually components of velocity vectors. *They may have positive or negative values determined by the direction of motion relative to the chosen coordinate origin.*

$$m_1 v_{1i} + m_2 v_{2i} = (m_1 + m_2)v_f \qquad (6.8)$$

In an **elastic head-on collision**, both momentum and kinetic energy are conserved.

$$m_1 v_{1i} + m_2 v_{2i} = m_1 v_{1f} + m_2 v_{2f} \qquad (6.10)$$

$$\tfrac{1}{2}m_1 v_{1i}^2 + \tfrac{1}{2}m_2 v_{2i}^2 = $$
$$\tfrac{1}{2}m_1 v_{1f}^2 + \tfrac{1}{2}m_2 v_{2f}^2 \qquad (6.11)$$

The **relative velocities** of two objects before and after an elastic collision are related as shown in Equation 6.14, which results from combining Equations 6.10 and 6.11 to eliminate m_1 and m_2. Equation 6.14 is valid only in elastic collisions.

$$v_{1i} - v_{2i} = -(v_{1f} - v_{2f}) \tag{6.14}$$

Consider a **two-dimensional elastic collision** in which an object m_1 moves along the x-axis and collides with m_2 initially at rest.

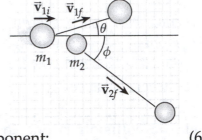

Momentum is conserved along each direction. Angles in Equations 6.15 and 6.16 are defined in the diagram.

x-component: $\tag{6.15}$
$$m_1 v_{1i} + 0 = m_1 v_{1f} \cos\theta + m_2 v_{2f} \cos\phi$$

y-component: $\tag{6.16}$
$$0 + 0 = m_1 v_{1f} \sin\theta - m_2 v_{2f} \sin\phi$$

For a **perfectly elastic collision** kinetic energy is conserved.

$$\tfrac{1}{2} m v_{1i}^2 + \tfrac{1}{2} m v_{2i}^2 = \tfrac{1}{2} m_{1f}^2 + \tfrac{1}{2} m_2 v_{2f}^2 \tag{6.17}$$

SUGGESTIONS, SKILLS, AND STRATEGIES

The following procedure is recommended when dealing with problems involving collisions between two objects:

1. Set up a coordinate system and define velocities with respect to that system. That is, objects moving in the direction selected as the positive direction of the x axis are considered as having a positive velocity and negative if moving in the negative x direction. It is convenient to have the x axis coincide with the direction of one of the initial velocities.

2. Show all velocity vectors with labels and include all the given information including scattering angles.

3. Write expressions for the momentum of each object before and after the collision. (In two-dimensional collision problems, write expressions for the x and y components of momentum before and after the collision.) Remember to include the appropriate signs for their velocity directions.

4. Now write expressions for the total momentum before and the total momentum after the collision and equate the two. (For two-dimensional collisions, separate expressions must be written for the momentum in the x and y directions. See Equations 6.15 and 6.16. *It is important to emphasize that it is the momentum of the system that is conserved, not the momentum of the individual objects.*

5. If the collision is perfectly inelastic (kinetic energy is not conserved and the two objects have a common velocity following the collision) you should then proceed to solve the momentum equations for the unknown quantities.

6. If the collision is elastic, kinetic energy is also conserved, so you can equate the total kinetic energy before the collision to the total kinetic energy after the collision. This gives an additional relationship between the various velocities. The conservation of kinetic energy for elastic collisions leads to the expression $v_{1i} - v_{2i} = -(v_{1f} - v_{2f})$, which is often easier to use in solving elastic collision problems than is an expression for conservation of kinetic energy.

REVIEW CHECKLIST

▷ The impulse of a force acting on a particle during some time interval equals the change in momentum of the particle, and the impulse equals the area under the force vs time graph.

▷ The momentum of any isolated system (or any system for which the net external force is zero) is conserved, regardless of the nature of the forces between the masses which comprise the system.

▷ There are two types of collisions that can occur between particles, namely elastic and inelastic collisions. Recognize that a perfectly inelastic collision is an inelastic collision in which the colliding particles stick together after the collision, and hence move as a composite particle. Kinetic energy is conserved in elastic collisions, but not in the case of inelastic collisions. Linear momentum is conserved in both elastic and inelastic collisions, if the net external force acting on the system of colliding objects is zero.

▷   The conservation of linear momentum applies not only to head-on collisions (one-dimensional), but also to glancing collisions (two- or three-dimensional). For example, in a two-dimensional collision, the total momentum in the x direction is conserved and the total momentum in the y direction is conserved.

▷   The equations for conservation of momentum and kinetic energy can be used to calculate the final velocities in a two-body head-on elastic collision. For a perfectly inelastic collision, the equation of conservation of momentum may be used to calculate the final velocity of the composite particle.

SOLUTIONS TO SELECTED END-OF-CHAPTER PROBLEMS

7.   A professional diver performs a dive from a platform 10 m above the water surface. Estimate the order of magnitude of the average impact force she experiences in her collision with the water. State the quantities you take as data and their values.

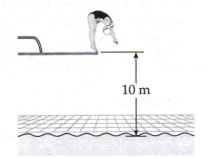

10 m

Solution

The speed just before impact with the water is found from conservation of mechanical energy,

$$\left(KE + PE_g\right)_f = \left(KE + PE_g\right)_i$$

If the diver starts from rest at height h above the water, this yields

$$\tfrac{1}{2}mv_{impact}^2 + 0 = 0 + mgh \qquad \text{or} \qquad v_{impact} = \sqrt{2gh}$$

As the diver is being brought to rest following impact with the water, she experiences an impulse given by

$$\vec{\mathbf{F}}_{av}\left(\Delta t\right) = \Delta\vec{\mathbf{p}} = m\vec{\mathbf{v}}_f - m\vec{\mathbf{v}}_i = 0 - m\vec{\mathbf{v}}_{impact}$$

Taking upward as the positive direction, the average force experienced by the diver during the interval, of duration Δt, between impact and the instant she comes to rest is

$$\vec{\mathbf{F}}_{av} = \frac{-m\vec{\mathbf{v}}_{impact}}{\Delta t} = \frac{-m\left(-\sqrt{2gh}\right)}{\Delta t} = \frac{m\sqrt{2gh}}{\Delta t} \quad \text{(upward)}$$

Assuming a mass of 55 kg and an impact time of $\Delta t \cong 1.0\text{ s}$, the magnitude of this average force is

$$F_{av} = \frac{\left(55\text{ kg}\right)\sqrt{2\left(9.80\text{ m/s}^2\right)\left(10\text{ m}\right)}}{1.0\text{ s}} = 770\text{ N} \quad \text{or} \quad F_{av} \sim 10^3\text{ N} \qquad \Diamond$$

13.   The forces shown in the force vs. time diagram in Figure P6.13 act on a 1.5-kg particle. Find (a) the impulse for the interval $t = 0$ to $t = 3.0$ s and (b) the impulse for the interval from $t = 0$ to $t = 5.0$ s. (c) If the forces act on a 1.5-kg particle that is initially at rest, find the particle's speed at $t = 3.0$ s and at $t = 5.0$ s.

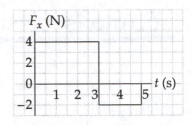

Figure P6.13

Solution

(a)   The impulse is the area under the curve between $t = 0$ and $t = 3.0$ s.

That is, $I = (4.0\ \text{N})(3.0\ \text{s}) = 12\ \text{N} \cdot \text{s}$ ◊

(b)   The area under the curve between $t = 0$ and $t = 5.0$ s is:

$$I = (4.0\ \text{N})(3.0\ \text{s}) + (-2.0\ \text{N})(2.0\ \text{s}) = 8.0\ \text{N} \cdot \text{s}$$ ◊

(c) $I = F_{av}(\Delta t) = \Delta p = m(v_f - v_i)$

so $v_f = v_i + \dfrac{I}{m}$

At 3.0 s: $v_f = v_i + \dfrac{I}{m} = 0 + \dfrac{12\ \text{N} \cdot \text{s}}{1.5\ \text{kg}} = 8.0\ \text{m/s}$ ◊

At 5.0 s: $v_f = v_i + \dfrac{I}{m} = 0 + \dfrac{8.0\ \text{N} \cdot \text{s}}{1.5\ \text{kg}} = 5.3\ \text{m/s}$ ◊

15. The front 1.20 m of a 1 400 kg car is designed as a "crumple zone" that collapses to absorb the shock of a collision. If a car traveling 25.0 m/s stops uniformly in 1.20 m, (a) how long does the collision last, (b) what is the magnitude of the average force on the car, and (c) what is the acceleration of the car? Express the acceleration as a multiple of the acceleration of gravity.

Solution

(a)   The duration of the collision is the time required for the car to travel the 1.20-m length of the "crumple zone". Assuming a uniform deceleration of the car, this time is given by

$$\Delta t = \frac{\Delta x}{v_{av}} = \frac{\Delta x}{\left(v_f + v_i\right)/2} = \frac{2(1.20 \text{ m})}{0 + 25.0 \text{ m/s}} = 9.60 \times 10^{-2} \text{ s} \qquad \Diamond$$

(b)   From the Impulse-momentum theorem, $\vec{\mathbf{I}} = \vec{\mathbf{F}}_{av}\Delta t = \Delta\vec{\mathbf{p}}$, the magnitude of the average force exerted on the car during the impact is

$$F_{av} = \frac{\left|\Delta\vec{\mathbf{p}}\right|}{\Delta t} = \frac{m\left|\vec{\mathbf{v}}_f - \vec{\mathbf{v}}_i\right|}{\Delta t} = \frac{(1\,400 \text{ kg})\left|0 - 25.0 \text{ m/s}\right|}{9.60 \times 10^{-2} \text{ s}} = 3.65 \times 10^5 \text{ N} \qquad \Diamond$$

(c)   The magnitude of the average acceleration of the car during this collision is

$$a_{av} = \frac{\left|\vec{\mathbf{v}}_f - \vec{\mathbf{v}}_i\right|}{\Delta t} = \frac{\left|0 - 25.0 \text{ m/s}\right|}{9.60 \times 10^{-2} \text{ s}} = 260 \text{ m/s}^2 = \left(260 \text{ m/s}^2\right)\left(\frac{1\,g}{9.80 \text{ m/s}^2}\right) = 26.6\,g \qquad \Diamond$$

21.   A 45.0-kg girl is standing on a 150-kg plank. The plank, originally at rest, is free to slide on a frozen lake, which is a flat, frictionless surface. The girl begins to walk along the plank at a constant velocity of 1.50 m/s to the right relative to the plank. (a) What is her velocity relative to the surface of the ice? (b) What is the velocity of the plank relative to the surface of the ice?

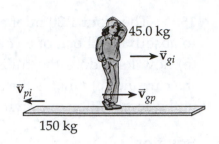

Solution

(a)   The velocity of the girl relative to the ice, $\vec{v}_{GI}$, is $\vec{v}_{GI} = \vec{v}_{GP} + \vec{v}_{PI}$

where $\vec{v}_{GP}$ = velocity of the girl relative to the plank

and $\vec{v}_{PI}$ = velocity of the plank relative to the ice.

Since we are given that $\vec{v}_{GP} = 1.50 \text{ m/s}$

this becomes $\vec{v}_{GI} = 1.50 \text{ m/s} + \vec{v}_{PI}$ **[1]**

Conservation of momentum gives $m_G \vec{v}_{GI} + m_P \vec{v}_{PI} = 0$

or $\vec{v}_{PI} = -\left(\dfrac{m_G}{m_P}\right)\vec{v}_{GI}$ **[2]**

Equation [1] becomes $\left(1 + \dfrac{m_G}{m_P}\right)\vec{v}_{GI} = +1.50 \text{ m/s}$

or $\vec{v}_{GI} = \dfrac{+1.50 \text{ m/s}}{1 + \left(\dfrac{45.0 \text{ kg}}{150 \text{ kg}}\right)} = +1.15 \text{ m/s}$ ◊

(b)   Then, using Equation [2], $\vec{v}_{PI} = -\left(\dfrac{45.0 \text{ kg}}{150 \text{ kg}}\right)(1.15 \text{ m/s}) = -0.346 \text{ m/s}$

or $\vec{v}_{PI} = 0.346 \text{ m/s}$ directed opposite to the girl's motion   ◊

30. An 8.00-g bullet is fired into a 250-g block that is initially at rest at the edge of a table of height 1.00 m (Fig. P6.30). The bullet remains in the block, and after the impact the block lands 2.00 m from the bottom of the table. Determine the initial speed of the bullet.

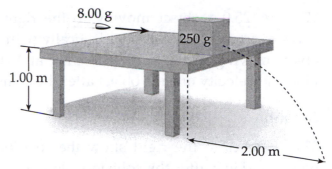

Figure P6.30

Solution

First, consider the motion of the block-bullet combination for the time interval from the instant just after the bullet embeds itself in the block until the block hits the floor. During this interval, the combination is a projectile with an initial velocity that is horizontal ($v_{0y} = 0$) with magnitude v_{0x}. The time the combination is in flight is found from $\Delta y = v_{0y}t + \frac{1}{2}a_y t^2$, which gives

$$-1.00 \text{ m} = 0 + \tfrac{1}{2}\left(-9.80 \text{ m/s}^2\right)t^2 \text{ or } t = \sqrt{\frac{2.00 \text{ m}}{9.80 \text{ m/s}^2}} = 0.452 \text{ s}$$

Taking the positive x direction to be the direction of the bullet's motion in Figure P6.30, the velocity of the block-bullet combination immediately after the bullet collides with the block is

$$V = v_{0x} = \frac{\Delta x}{t} = \frac{2.00 \text{ m}}{0.452 \text{ s}} = 4.43 \text{ m/s}$$

Now apply conservation of momentum from before to just after the bullet-block collision:

$$p_i = p_f \quad \Rightarrow \quad \left(p_{bullet}\right)_i + \left(p_{block}\right)_i = \left(p_{combination}\right)_f$$

or $\qquad mv_{bullet} + M(0) = (m + M)V$

giving $v_{bullet} = \left(1 + \dfrac{M}{m}\right)V = \left(1 + \dfrac{250 \text{ g}}{8.00 \text{ g}}\right)(4.43 \text{ m/s}) = 143 \text{ m/s}$ ◊

Observe that $v_{bullet} = (143 \text{ m/s})\left(\dfrac{1 \text{ mi/h}}{0.447 \text{ m/s}}\right) \approx 320 \text{ mph}$

37.   A 25.0-g object moving to the right at 20.0 cm/s overtakes and collides elastically with a 10.0-g object moving in the same direction at 15.0 cm/s. Find the velocity of each object after the collision.

Solution

The sketches to the right show the conditions just before and just after the collision. Note the choice of the $+x$ direction.

Applying conservation of momentum to this collision gives $\vec{p}_{1f} + \vec{p}_{2f} = \vec{p}_{1i} + \vec{p}_{2i}$:

$$(25.0 \text{ g})\vec{v}_{1f} + (10.0 \text{ g})\vec{v}_{2f} = (25.0 \text{ g})(+20.0 \text{ cm/s}) + (10.0 \text{ g})(+15.0 \text{ cm/s})$$

which reduces to: $2.50\vec{v}_{1f} + \vec{v}_{2f} = +65.0 \text{ cm/s}$ [1]

This is a perfectly elastic, head-on collision,

so $\vec{v}_{1i} - \vec{v}_{2i} = -\left(\vec{v}_{1f} - \vec{v}_{2f}\right)$

or $+20.0 \text{ cm/s} - (+15.0 \text{ cm/s}) = -\vec{v}_{1f} + \vec{v}_{2f}$

This reduces to : $\vec{v}_{2f} = \vec{v}_{1f} + 5.00 \text{ cm/s}$ [2]

Substitution of Equation [2] into Equation [1] yields $3.50\vec{v}_{1f} = +60.0 \text{ cm/s}$

Thus, $\vec{v}_{1f} = +17.1 \text{ cm/s} = 17.1 \text{ cm/s in the } +x \text{ direction}$ ◊

Then, Equation [2] gives: $\vec{v}_{2f} = +22.1 \text{ cm/s} = 22.1 \text{ cm/s in } +x \text{ direction}$ ◊

41. A 90-kg fullback moving east with a speed of 5.0 m/s is tackled by a 95-kg opponent running north at 3.0 m/s. If the collision is perfectly inelastic, calculate (a) the velocity of the players just after the tackle and (b) the kinetic energy lost as a result of the collision. Can you account for the missing energy?

Solution

(a) Choose east to be the positive x-direction and north the positive y-direction. Conservation of momentum during the tackle gives:

$$\vec{P}_f = \vec{P}_i \quad \Rightarrow \quad \sum(p_x)_f = \sum(p_x)_i \quad \text{and} \quad \sum(p_y)_f = \sum(p_y)_i$$

Conserving momentum in the x-direction first:

$$\left(m_{back} + m_{opponent}\right)v_{fx} = m_{back}(v_1)_x + m_{opponent}(v_2)_x$$

or $\quad (90 \text{ kg} + 95 \text{ kg})v_{fx} = (90 \text{ kg})(+5.0 \text{ m/s}) + (95 \text{ kg})(0)$

$$v_{fx} = \frac{+450 \text{ kg} \cdot \text{m/s}}{185 \text{ kg}}$$

Then, conserving momentum in the y-direction:

$$\left(m_{back} + m_{opponent}\right)v_{fy} = m_{back}(v_1)_y + m_{opponent}(v_2)_y$$

or $\quad (90 \text{ kg} + 95 \text{ kg})v_{fy} = (90 \text{ kg})(0) + (95 \text{ kg})(+3.0 \text{ m/s})$

$$v_{fy} = \frac{+285 \text{ kg} \cdot \text{m/s}}{185 \text{ kg}}$$

The magnitude and direction of the final velocity are:

$$v_f = \sqrt{\left(v_{fx}\right)^2 + \left(v_{fy}\right)^2} = \frac{\sqrt{(450)^2 + (285)^2} \text{ kg} \cdot \text{m/s}}{185 \text{ kg}} = 2.9 \text{ m/s}$$

and $\quad \theta = \tan^{-1}\left(\dfrac{v_{fy}}{v_{fx}}\right) = \tan^{-1}\left(\dfrac{285}{450}\right) = 32°$

Hence, $\qquad\qquad\qquad \vec{v}_f = 2.9 \text{ m/s at } 32° \text{ north of east}$ ◇

(b)   The kinetic energy converted into other forms of energy during the collision is

$$KE_{lost} = KE_i - KE_f = \left(\tfrac{1}{2}m_{back}v_1^2 + \tfrac{1}{2}m_{opponent}v_2^2\right) - \tfrac{1}{2}\left(m_{back} + m_{opponent}\right)v_f^2$$

$$= \tfrac{1}{2}(90 \text{ kg})(5.0 \text{ m/s})^2 + \tfrac{1}{2}(95 \text{ kg})(3.0 \text{ m/s})^2 - \tfrac{1}{2}(185 \text{ kg})(2.9 \text{ m/s})^2$$

or $KE_{lost} = 7.9 \times 10^2 \text{ J}$ ◊

45. A billiard ball moving at 5.00 m/s strikes a stationary ball of the same mass. After the collision, the first ball moves at 4.33 m/s at an angle of 30° with respect to the original line of motion. (a) Find the velocity (magnitude and direction) of the second ball after collision. (b) Was this an inelastic collision or an elastic collision?

Solution

The sketch given at the right shows this glancing collision both just before impact and just after impact. The resultant momentum vector after collision is the same as the resultant momentum vector before collision, so

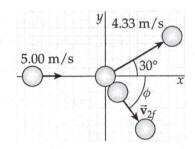

$$\left(\sum p_x\right)_f = \left(\sum p_x\right)_i \text{ and } \left(\sum p_y\right)_f = \left(\sum p_y\right)_i$$

(a)   First, looking at the x-components of momentum, this gives

$$m(4.33 \text{ m/s})\cos 30.0° + mv_{2f} \cos\phi = m(5.00 \text{ m/s}) + 0$$

Thus $3.75 \text{ m/s} + v_{2f} \cos\phi = 5.00 \text{ m/s}$

and $v_{2f} \cos\phi = 1.25 \text{ m/s}$ [1]

Then, considering the y-components of momentum gives

$$m(4.33 \text{ m/s})\sin 30.0° + mv_{2f} \sin\phi = 0 + 0$$

which reduces to $v_{2f} \sin\phi = -2.17 \text{ m/s}$ [2]

Squaring Equations [1] and [2] and adding the results yields

$$v_{2f}^2\left(\cos^2\phi + \sin^2\phi\right) = (1.25 \text{ m/s})^2 + (-2.17 \text{ m/s})^2$$

Since (by the trigonometric identity) $\cos^2\phi + \sin^2\phi = 1$, the speed of the second ball after the collision is found as:

$$v_{2f}^2 = 6.25 \text{ m}^2/\text{s}^2 \quad \text{and} \quad v_{2f} = 2.50 \text{ m/s} \qquad \Diamond$$

Since we were asked for the velocity, we also need the angle ϕ of the velocity.

Dividing Equation [2] by Equation [1] gives $\dfrac{\sin\phi}{\cos\phi} = \dfrac{-2.17 \text{ m/s}}{1.25 \text{ m/s}}$

Since $\tan\phi = \dfrac{\sin\phi}{\cos\phi}$, $\tan\phi = \dfrac{-2.17 \text{ m/s}}{1.25 \text{ m/s}}$

which yields two answers:

$$\phi = \tan^{-1}\left(\dfrac{-2.17}{1.25}\right) = -60°, \ 120°$$

Looking at the original diagram, we can see that conservation of momentum would not allow the answer $\phi = +120°$. If $\phi = 120°$ counterclockwise from the $+x$-axis, $\left(\Sigma p_y\right)_f$ cannot be zero as required. Therefore, we know that only the first angle is correct; the velocity of the second ball is 2.50 m/s at 60° below the $+x$-axis. ◊

(b)   The total kinetic energy before collision is

$$KE_i = KE_{1i} + KE_{2i} \quad \text{or} \quad KE_i = \tfrac{1}{2}m(5.00 \text{ m/s})^2 + 0 = m\left(12.5 \text{ m}^2/\text{s}^2\right)$$

The total kinetic energy after collision is

$$KE_f = KE_{1f} + KE_{2f} = \tfrac{1}{2}m(4.33 \text{ m/s})^2 + \tfrac{1}{2}m(2.50 \text{ m/s})^2$$

which yields $KE_f = m\left(12.5 \text{ m}^2/\text{s}^2\right)$

Since the total kinetic energy after collision is the same as that before collision, **the collision is elastic.** ◊

51.   A 2.0-g particle moving at 8.0 m/s makes a perfectly elastic head-on collision with a resting 1.0-g object. (a) Find the speed of each particle after the collision. (b) Find the speed of each particle after the collision if the stationary particle has a mass of 10 g. (c) Find the final kinetic energy of the incident 2.0-g particle in the situations described in (a) and (b). In which case does the incident particle lose more kinetic energy?

Solution

Note that the initial velocity of the target object is zero ($v_{2i} = 0$).

From conservation of momentum, $m_1 v_{1f} + m_2 v_{2f} = m_1 v_{1i} + 0$ [1]

For head-on elastic collisions, $v_{1i} - 0 = -\left(v_{1f} - v_{2f}\right) \Rightarrow v_{2f} = v_{1f} + v_{1i}$ [2]

Substituting Equation (2) into Equation (1) and solving for v_{1f} gives

$$v_{1f} = \left(\frac{m_1 - m_2}{m_1 + m_2}\right)v_{1i}$$

Substituting this result into Equation (2) and simplifying yields

$$v_{2f} = \left(\frac{2m_1}{m_1 + m_2}\right)v_{1i}$$

(a)   If $m_1 = 2.0$ g, $m_2 = 1.0$ g, and $v_{1i} = 8.0$ m/s, then

$$v_{1f} = \frac{8}{3} \text{ m/s and } v_{2f} = \frac{32}{3} \text{ m/s}$$ ◇

(b)   If $m_1 = 2.0$ g, $m_2 = 10$ g, and $v_{1i} = 8.0$ m/s, we find

$$v_{1f} = -\frac{16}{3} \text{ m/s and } v_{2f} = \frac{8}{3} \text{ m/s}$$ ◇

(c)   The final kinetic energy of the 2.0 g particle in each case is:

Case (a): $KE_{1f} = \frac{1}{2}m_1 v_{1f}^2 = \frac{1}{2}\left(2.0 \times 10^{-3} \text{ kg}\right)\left(\frac{8}{3} \text{ m/s}\right)^2 = 7.1 \times 10^{-3}$ J   ◇

Case (b): $KE_{1f} = \frac{1}{2}m_1 v_{1f}^2 = \frac{1}{2}\left(2.0 \times 10^{-3} \text{ kg}\right)\left(-\frac{16}{3} \text{ m/s}\right)^2 = 2.8 \times 10^{-2}$ J   ◇

Since the incident kinetic energy is the same in cases (a) and (b), we observe that the incident particle loses more kinetic energy in case (a).   ◇

58.   Tarzan, whose mass is 80.0 kg, swings from a 3.00-m vine that is horizontal when he starts. At the bottom of his arc, he picks up 60.0-kg Jane in a perfectly inelastic collision. What is the height of the highest tree limb they can reach on their upward swing?

Solution

Many students will attempt to use conservation of energy from the instant Tarzan starts his downward swing until he and Jane come to rest at the highest point on the upward swing. This approach is incorrect because a significant portion of the original mechanical energy is converted into other forms of energy during the inelastic collision as Tarzan picks up Jane. The correct approach is similar to that used in Example 6.5 of the textbook.

The method of the following solution is to use energy conservation from the start of Tarzan's swing until just before he "collides" with Jane. Then, conservation of momentum is used to get through the collision. Finally, conservation of energy can be used from immediately after the collision until they reach the highest point of their upward swing.

Energy conservation for Tarzan's downward swing:

$$\left(KE+PE_g\right)_f=\left(KE+PE_g\right)_i \Rightarrow \tfrac{1}{2}m_T v_T^2+0=0+m_T gy_i \quad \text{or} \quad v_T=\sqrt{2gy_i}$$

or Tarzan's speed just before contacting Jane is

$$v_T=\sqrt{2\left(9.80\ \text{m/s}^2\right)\left(3.00\ \text{m}\right)}=7.67\ \text{m/s}$$

Conservation of momentum during the collision:

$$p_f=p_i \Rightarrow \left(m_T+m_J\right)V=m_T v_T+m_J\left(0\right)$$

so the speed of Tarzan and Jane, moving as a single unit, as they start the upward swing is

$$V=\frac{m_T v_T}{m_T+m_J}=\frac{\left(80.0\ \text{kg}\right)\left(7.67\ \text{m/s}\right)}{80.0\ \text{kg}+60.0\ \text{kg}}=4.38\ \text{m/s}$$

Conservation of energy during the upward swing:

$$\left(KE+PE_g\right)_f=\left(KE+PE_g\right)_i \Rightarrow 0+\left(m_T+m_J\right)gy_f=\tfrac{1}{2}\left(m_T+m_J\right)V^2+0 \quad \text{or} \quad y_f=\frac{V^2}{2g}$$

so the height of the highest tree limb they can reach on the upward swing is

$$y_f=\frac{V^2}{2g}=\frac{\left(4.38\ \text{m/s}\right)^2}{2\left(9.80\ \text{m/s}^2\right)}=0.980\ \text{m}$$

◊

63. A neutron in a reactor makes an elastic collision head on with a carbon atom that is initially at rest. (The mass of the carbon nucleus is about 12 times that of the neutron.) (a) What fraction of the neutron's kinetic energy is transferred to the carbon nucleus? (b) If the neutron's initial kinetic energy is 1.6×10^{-13} J, find its final kinetic energy and the kinetic energy of the carbon nucleus after the collision.

Solution

Let particle 1 be the neutron and particle 2 be the carbon nucleus. Then, we are given that $m_2 = 12\,m_1$.

(a) From conservation of momentum

$$m_2 v_{2f} + m_1 v_{1f} = m_1 v_{1i} + 0$$

Since $m_2 = 12\,m_1$, this reduces to

$$12 v_{2f} + v_{1f} = v_{1i} \qquad [1]$$

For a head-on elastic collision,

$$v_{1i} - v_{2i} = -\left(v_{1f} - v_{2f}\right)$$

Since $v_{2i} = 0$, this becomes

$$v_{2f} - v_{1f} = v_{1i} \qquad [2]$$

Adding Equations [1] and [2] yields

$$13 v_{2f} = 2 v_{1i}$$

or

$$v_{2f} = \frac{2}{13} v_{1i}$$

The initial kinetic energy of the neutron is $KE_{1i} = \frac{1}{2} m_1 v_{1i}^2$, and the final kinetic energy of the carbon nucleus is

$$KE_{2f} = \frac{1}{2} m_2 v_{2f}^2 = \frac{1}{2}(12\,m_1)\left(\frac{4}{169} v_{1i}^2\right) = \frac{48}{169}\left(\frac{1}{2} m_1 v_{1i}^2\right) = \frac{48}{169} KE_{1i}$$

The fraction of kinetic energy transferred is $\quad \dfrac{KE_{2f}}{KE_{1i}} = \dfrac{48}{169} = 0.28 \qquad \Diamond$

(b) If $\quad KE_{1i} = 1.6 \times 10^{-13}$ J

then $\quad KE_{2f} = \dfrac{48}{169} KE_{1i} = \dfrac{48}{169}\left(1.6 \times 10^{-13}\text{ J}\right) = 4.5 \times 10^{-14}\text{ J} \qquad \Diamond$

The kinetic energy remaining with the neutron is

$$KE_{1f} = KE_{1i} - KE_{2f} = 1.6 \times 10^{-13}\text{ J} - 4.5 \times 10^{-14}\text{ J} = 1.1 \times 10^{-13}\text{ J} \qquad \Diamond$$

69.   A tennis ball of mass 57.0 g is held just above a basketball of mass 590 g. With their centers vertically aligned, both balls are released from rest at the same time, to fall through a distance of 1.20 m, as shown in Figure P6.69. (a) Find the magnitude of the downward velocity with which the basketball reaches the ground. Assume that an elastic collision with the ground instantaneously reverses the velocity of the basketball while the tennis ball is still moving down. Next, the two balls meet in an elastic collision. (b) To what height does the tennis ball rebound?

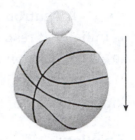

Solution

(a)   When released, both balls start from rest and fall freely under the influence of gravity for 1.20 m before the basketball touches the ground. Their downward speed just before the basketball hits ground may be found from $v_y^2 = v_{0y}^2 + 2a_y \Delta y$. This yields:

$$v_i = \sqrt{v_{0y}^2 + 2a_y \Delta y} = \sqrt{0 + 2(-9.80 \text{ m/s}^2)(-1.20 \text{ m})} = 4.85 \text{ m/s}$$ ◊

(b)   Immediately after the basketball rebounds from the floor, it and the tennis ball meet in an elastic collision. The vertical velocity of each ball just before the collision is:

    For the tennis ball: $v_{1i} = -v_i = -4.85 \text{ m/s}$

    For the basketball: $v_{2i} = +v_i = +4.85 \text{ m/s}$

The velocity of the tennis ball immediately after this elastic collision may be found as follows:

Conservation of momentum gives

$$(57.0 \text{ g})v_{1f} + (590 \text{ g})v_{2f} = (57.0 \text{ g})(-4.85 \text{ m/s}) + (590 \text{ g})(+4.85 \text{ m/s})$$

which reduces to $v_{2f} = 4.38 \text{ m/s} - (9.66 \times 10^{-2})v_{1f}$ [1]

From the characteristics of a head-on elastic collision, we know that

$$v_{1f} - v_{2f} = -(v_{1i} - v_{2i}) = -(-4.85 \text{ m/s} - 4.85 \text{ m/s}) = +9.70 \text{ m/s}$$

or $v_{1f} = 9.70 \text{ m/s} + v_{2f}$ [2]

Substituting Equation (1) into Equation (2) and simplifying gives

$$(1+9.66\times10^{-2})v_{1f} = 9.70 \ \text{m/s} + 4.38 \ \text{m/s}$$

or $\qquad v_{1f} = \dfrac{14.1 \ \text{m/s}}{1.10} = +12.8 \ \text{m/s}$

The vertical displacement of the tennis ball during its rebound is then found from $v_y^2 = v_{0y}^2 + 2a_y\Delta y$ to be

$$\Delta y = \frac{v_y^2 - v_{0y}^2}{2a_y} = \frac{0-(12.8 \ \text{m/s})^2}{2(-9.80 \ \text{m/s}^2)} = 8.41 \ \text{m} \qquad\qquad \lozenge$$

Chapter 7
Rotational Motion and the Law of Gravity

NOTES FROM SELECTED CHAPTER SECTIONS

7.1 Angular Speed and Angular Acceleration

Pure rotational motion refers to the motion of a rigid body about a fixed axis. In the case of **rotation about a fixed axis,** every particle on the rigid body has the same angular velocity and the same angular acceleration.

One **radian** (rad) is the angle subtended by an arc length equal to the radius of the arc. That is, the angle measured in radians is given by the arc length divided by the corresponding radius. One complete rotation of 360 degrees equals 2π radians.

7.2 Rotational Motion Under Constant Angular Acceleration

The equations for rotational motion under constant angular acceleration are of the same form as those for linear motion under constant linear acceleration with the substitutions $x \rightarrow \theta$, $v \rightarrow \omega$, and $a \rightarrow \alpha$.

7.3 Relations Between Angular and Linear Quantities

When a **rigid body rotates about a fixed axis,** every point in the object moves along a circular path which has its center at the axis of rotation. The instantaneous velocity of each point is directed along a tangent to the circle. Every point on the object experiences the same angular speed; however, points that are different distances from the axis of rotation have different tangential speeds. The value of each of the linear quantities [displacement (s), velocity (v), and acceleration (a_t)] is equal to the radial distance from the axis multiplied by the corresponding angular quantity, θ, ω, and α.

7.4   Centripetal Acceleration

In circular motion, the centripetal acceleration is directed inward toward the center of the circle and has a magnitude given either by v^2/r or $r\omega^2$. The force that causes centripetal acceleration acts toward the center of the circular path along which the object moves. If the force vanishes, the object does not continue to move in its circular path; instead, it moves along a straight-line path tangent to the circle.

7.5   Newtonion Gravitation

There are several important features of the law of universal gravitation.

The gravitational force:

- is an action-at-a-distance force that always exists between two particles regardless of the medium that separates them

- varies as the inverse square of the distance between the particles and therefore decreases rapidly with increasing separation, and

- is proportional to the product of their masses.

The gravitational force exerted by a uniform spherical mass on a particle outside the sphere is the same as if the entire mass of the sphere were concentrated at the center.

7.6   Kepler's Laws

Kepler's laws applied to the solar system are:

1.   All planets move in elliptical orbits with the Sun at one of the focal points.

2.   A line drawn from the Sun to any planet sweeps out equal areas in equal time intervals.

3.   The square of the orbital period of any planet is proportional to the cube of the average distance from the planet to the Sun.

EQUATIONS AND CONCEPTS

Arc length, s is the distance traveled by a particle as it moves along a circular path of radius r. The radial line from the center of the path to the particle sweeps out an angle, θ.

$$\theta \equiv \frac{s}{r} \qquad (7.1)$$

where $\theta \, (\text{rad}) = \left(\frac{\pi}{180°}\right)(\theta \, (\text{deg}))$

The **radian**, a unit of angular measure, is the ratio of two lengths (arc length to radius) and hence is a dimensionless quantity

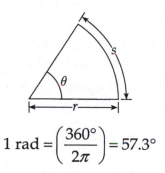

$$1 \text{ rad} = \left(\frac{360°}{2\pi}\right) = 57.3°$$

The **average angular velocity** of a rotating object is the ratio of the angular displacement to the time interval during which the angular displacement occurs.

$$\omega_{av} \equiv \frac{\theta_f - \theta_o}{t_f - t_o} = \frac{\Delta\theta}{\Delta t} \qquad (7.3)$$

The **average angular acceleration** of a rotating object is the ratio of change in angular velocity to the time interval during which the change in velocity occurs.

$$\alpha_{av} \equiv \frac{\omega_f - \omega_o}{t_f - t_i} = \frac{\Delta\omega}{\Delta t} \qquad (7.5)$$

To the right are given the equations of **rotational motion** with constant acceleration and the corresponding equations for **linear motion** with constant acceleration. Note that the rotational equations, involving the angular variables $\Delta\theta$, ω, and α, have a one-to-one correspondence with the equations of linear motion, involving the variables Δx, v, and a.

$$\omega = \omega_o + \alpha t \qquad (7.7)$$

$$v = v_o + at$$

$$\Delta\theta = \omega_o t + \tfrac{1}{2}\alpha t^2 \qquad (7.8)$$

$$\Delta x = v_o t + \tfrac{1}{2}at^2$$

$$\omega^2 = \omega_o^2 + 2\alpha\,\Delta\theta \qquad (7.9)$$

$$v^2 = v_o^2 + 2a\Delta x$$

The **tangential velocity** and **tangential acceleration** of a given point on a rotating object are related to the corresponding angular quantities via the radius of the circular path along which the point moves. The tangential velocity and tangential acceleration are directed along the tangent to the circular path (and therefore perpendicular to the radius from the center of rotation). *Every point on a rotating object has the same value of ω and the same value of α. However, points which are at different distances from the axis of rotation have different values of v_t and a_t.*

$$v_t = r\omega \tag{7.10}$$

$$a_t = r\alpha \tag{7.11}$$

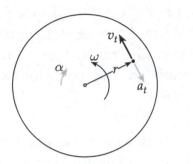

In the diagram above, a rigid disk is rotating counterclockwise ($+\omega$) about an axis through its center with a decreasing rate of rotation ($-\alpha$)

An **object in circular motion** has a centripetal acceleration directed toward the center of the circle with a magnitude which depends on the values of the tangential velocity and the radius of the path.

$$a_c = \frac{v^2}{r} \tag{7.13}$$

or

$$a_c = r\omega^2 \tag{7.17}$$

The **total acceleration** of an object which is moving along a circular path with increasing or decreasing speed has both a tangential component of acceleration (tangent to the instantaneous direction of travel) and a centripetal component of acceleration (perpendicular to the instantaneous direction of travel). The magnitude and direction of the total acceleration can be found by the usual methods of vector addition.

$$a = \sqrt{a_t^2 + a_c^2} \tag{7.18}$$

$$\theta = \tan^{-1}\left(\frac{a_t}{a_c}\right)$$

The resultant force that maintains motion along a circular path is always directed toward the center of the path. This force is also called a radial force.

$$F_r = ma_c = m\frac{v^2}{r} \tag{7.19}$$

The **law of universal gravitation** states that every particle in the universe attracts every other particle with a force that is directly proportional to the product of their masses and inversely proportional to the square of the distance between them. The constant G is called the universal gravitational constant.

$$F = G\frac{m_1 m_2}{r^2} \qquad (7.20)$$

$$G = 6.673 \times 10^{-11} \ \text{N} \cdot \text{m}^2/\text{kg}^2$$

The **gravitational force** between two masses m_1 and m_2 is one of attraction; each mass exerts a force of attraction on the other. These two forces form an action-reaction pair. *The magnitudes of the forces of gravitational attraction on the two masses are equal regardless of the relative values of m_1 and m_2.* This is in agreement with Newton's third law (action - reaction).

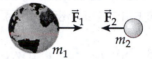

The gravitational force is an attractive force

The **escape speed** of an object projected upward from the Earth's surface is independent of the mass of the object.

$$v_{esc} = \sqrt{\frac{2GM_E}{R_E}} \qquad (7.22)$$

Kepler's third law states that the square of the orbital period of a planet is proportional to the cube of the mean distance from the planet to the Sun. For an Earth satellite, M_S in Equation 7.23 must be replaced by M_E, the mass of the Earth.

$$T^2 = \left(\frac{4\pi^2}{GM_S}\right)r^3 = K_S r^3 \qquad (7.23)$$

The value of K_S is independent of the mass of the planet.

$$K_S = \frac{4\pi^2}{GM_S} = 2.97 \times 10^{-19} \ \text{s}^2/\text{m}^3$$

SUGGESTIONS, SKILLS, AND STRATEGIES

When solving problems involving rotational motions and centripetal accelerations, the following points should be kept in mind:

1. Draw a free-body diagram of the object(s) under consideration, showing all forces that act on it (them).

2. Choose a coordinate system with one axis tangent to the path followed by the object and the other axis perpendicular to the plane of the circular path.

3. Find the net force toward the center of the circular path. This is the centripetal force, the force which causes the centripetal acceleration.

4. From this point onward, the steps are virtually identical to those encountered when solving Newton's second law problems with $F_x = ma_x$ and $F_y = ma_y$. In this case, Newton's second law is applied along the radial (directed toward the center) and tangential directions. Also, you should note that the magnitude of the centripetal acceleration can always be written as $a_c = v^2/r$.

REVIEW CHECKLIST

▷ Quantitatively, the angular displacement, angular velocity, and angular acceleration for a rigid body system in rotational motion are related to the distance traveled, tangential velocity, and tangential acceleration, respectively. Each linear quantity is calculated by multiplying the corresponding angular quantity by the radius for that particular object or point.

▷ If a body rotates about a fixed axis, every particle on the body has the same angular velocity and angular acceleration. For this reason, rotational motion can be simply described using these quantities. The formulas which describe angular motion are analogous to the corresponding set of equations pertaining to linear motion.

▷ A particle moving along a circular path with constant speed experiences an acceleration although the magnitude of the velocity remains constant. This is true because the direction of the velocity (always tangent to the circular path) changes with time. The direction of the acceleration is toward the center of the circular path.

▷ When both the magnitude and direction of $\vec{v}$ are changing with time, there are two components of acceleration for a particle moving on a curved path. In this case, the particle has a tangential component of acceleration and a radial component of acceleration.

▷ Newton's law of universal gravitation is an example of an inverse-square law, and it describes an attractive force between two particles separated by a distance r.

SOLUTIONS TO SELECTED END-OF-CHAPTER PROBLEMS

3. Find the angular speed of Earth about the Sun in radians per second and degrees per day.

Solution

The Earth completes a full orbit (360° or 2π radians) around the Sun in one year (3.156×10^7 s). Therefore, the average angular speed of the Earth's orbital motion is

$$\omega_{av} = \frac{\Delta\theta}{\Delta t} = \frac{2\pi \text{ rad}}{3.156\times10^7 \text{ s}} = 1.99\times10^{-7} \text{ rad/s} \qquad \Diamond$$

Converting to units of degrees per day yields

$$\omega_{av} = \left(1.99\times10^{-7} \text{ rad/s}\right)\left(\frac{57.3 \text{ deg}}{1 \text{ rad}}\right)\left(\frac{8.64\times10^4 \text{ s}}{1 \text{ day}}\right) = 0.986 \text{ deg/day} \qquad \Diamond$$

---

9. The diameters of the main rotor and tail rotor of a single-engine helicopter are 7.60 m and 1.02 m, respectively. The respective rotational speeds are 450 rev/min and 4 138 rev/min. Calculate the speeds of the tips of both rotors. Compare these speeds with the speed of sound, 343 m/s.

Solution

The tips of one of the rotors move in a circular path with the specified angular velocity (or rotational speed), ω, for that rotor. The linear speed of these rotor tips is given by $v_t = r\omega$, where r is the radius of the circular path.

For the main rotor: $\quad v_t = r\omega = \left(\frac{7.60 \text{ m}}{2}\right)\left(450 \frac{\text{rev}}{\text{min}}\right)\left(\frac{1 \text{ min}}{60 \text{ s}}\right)\left(\frac{2\pi \text{ rad}}{1 \text{ rev}}\right) = 179 \text{ m/s} \qquad \Diamond$

and $\quad v_t = \left(179 \text{ m/s}\right)\left(\frac{v_{sound}}{343 \text{ m/s}}\right) = 0.522\, v_{sound} \qquad \Diamond$

For the tail rotor: $\quad v_t = r\omega = \left(\frac{1.02 \text{ m}}{2}\right)\left(4\,138 \frac{\text{rev}}{\text{min}}\right)\left(\frac{1 \text{ min}}{60 \text{ s}}\right)\left(\frac{2\pi \text{ rad}}{1 \text{ rev}}\right) = 221 \text{ m/s} \qquad \Diamond$

and $\quad v_t = \left(221 \text{ m/s}\right)\left(\frac{v_{sound}}{343 \text{ m/s}}\right) = 0.644\, v_{sound} \qquad \Diamond$

13.   A rotating wheel requires 3.00 s to rotate 37.0 revolutions. Its angular velocity at the end of the 3.00-s interval is 98.0 rad/s. What is the constant angular acceleration of the wheel?

Solution

For the 3.00-s time interval of interest, we have the following information about the motion of the wheel:

$$\text{Angular displacement undergone} - \quad \Delta\theta = (37.0 \text{ rev})\left(\frac{2\pi \text{ rad}}{1 \text{ rev}}\right) = 74.0\pi \text{ rad}$$

$$\text{Final angular velocity} - \qquad\qquad \omega = 98.0 \text{ rad/s}$$

$$\text{Elapsed time} - \qquad\qquad t = 3.00 \text{ s}$$

The unknown quantities about the motion are:

$$\text{Initial angular velocity} - \qquad\qquad \omega_0$$

$$\text{Constant angular acceleration} - \qquad \alpha$$

From the relation $\omega = \omega_0 + \alpha t$, we obtain: $\qquad\qquad \omega_0 = \omega - \alpha t$

Substituting this into the relation $\Delta\theta = \omega_0 t + \frac{1}{2}\alpha t^2$ yields

$$\Delta\theta = (\omega - \alpha t)t + \tfrac{1}{2}\alpha t^2 = \omega t - \tfrac{1}{2}\alpha t^2$$

or $\qquad\qquad \alpha = \dfrac{2(\omega t - \Delta\theta)}{t^2}$

Thus, the angular acceleration of the wheel is

$$\alpha = \frac{2\left[(98.0 \text{ rad/s})(3.00 \text{ s}) - 74.0\pi \text{ rad}\right]}{(3.00 \text{ s})^2} = 13.7 \text{ rad/s}^2 \qquad\qquad \Diamond$$

If desired, the initial angular velocity of the wheel can now be found:

$$\omega_0 = \omega - \alpha t = 98.0 \text{ rad/s} - (13.7 \text{ rad/s}^2)(3.00 \text{ s}) = 57.0 \text{ rad/s}$$

17.   (a) What is the tangential acceleration of a bug on the rim of a 10-in.-diameter disk if the disk moves from rest to an angular speed of 78 rev/min in 3.0 s? (b) When the disk is at its final speed, what is the tangential velocity of the bug? (c) One second after the bug starts from rest, what are its tangential acceleration, centripetal acceleration, and total acceleration?

Solution

The angular velocity of the disk 3.0 s after starting from rest is:

$$\omega = 78 \ \frac{\text{rev}}{\text{min}} \left(\frac{2\pi \ \text{rad}}{1 \ \text{rev}} \right) \left(\frac{1 \ \text{min}}{60 \ \text{s}} \right) = 8.2 \ \text{rad/s}$$

and the bug follows a circular path of radius

$$r = 5.0 \ \text{in} \left(\frac{1 \ \text{m}}{39.37 \ \text{in}} \right) = 0.13 \ \text{m}$$

(a)   The constant angular acceleration of the disk is

$$\alpha = \frac{\omega - \omega_0}{t} = \frac{8.2 \ \text{rad/s} - 0}{3.0 \ \text{s}} = 2.7 \ \text{rad/s}^2$$

The tangential acceleration of the bug is $a_t = r\alpha$. Thus,

$$a_t = (0.13 \ \text{m})(2.7 \ \text{rad/s}^2) = 0.35 \ \text{m/s}^2 \qquad \Diamond$$

(b)   When the disk is rotating at its final angular velocity, $\omega = 8.2 \ \text{rad/s}$, and

$$v_t = r\omega = (0.13 \ \text{m})(8.2 \ \text{rad/s}) = 1.0 \ \text{m/s} \qquad \Diamond$$

(c)   Since both r and α are constant, the tangential acceleration, $a_t = r\alpha$, is also constant. Thus, at $t = 1.0 \ \text{s}$, $a_t = 0.35 \ \text{m/s}^2$ $\qquad \Diamond$

At $t = 1.0 \ \text{s}$, the tangential velocity of the bug is

$$v_t = (v_t)_{t=0} + a_t t = 0 + (0.35 \ \text{m/s}^2)(1.0 \ \text{s}) = 0.35 \ \text{m/s}$$

and the centripetal acceleration is

$$a_c = \frac{v_t^2}{r} = \frac{(0.35 \ \text{m/s})^2}{0.13 \ \text{m}} = 0.94 \ \text{m/s}^2 \qquad \Diamond$$

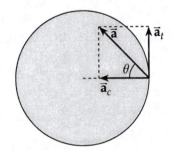

At $t = 1.0$ s , the total acceleration has magnitude

$$a = \sqrt{a_c^2 + a_t^2} = \sqrt{(0.94 \text{ m/s}^2)^2 + (0.35 \text{ m/s}^2)^2} = 1.0 \text{ m/s}^2 \qquad \Diamond$$

at $\quad \theta = \tan^{-1}\left(\dfrac{a_t}{a_c}\right) = \tan^{-1}\left(\dfrac{0.35 \text{ m/s}^2}{0.94 \text{ m/s}^2}\right) = 20° \qquad \Diamond$

23. A 50.0-kg child stands at the rim of a merry-go-round of radius 2.00 m, rotating with an angular speed of 3.00 rad/s. (a) What is the child's centripetal acceleration? (b) What is the minimum force between her feet and the floor of the carousel that is required to keep her in the circular path? (c) What minimum coefficient of static friction is required? Is the answer you found reasonable? In other words, is she likely to stay on the merry-go-round?

Solution

(a)   When the child stands, without slipping, at the rim of the rotating merry-go-round, she follows a horizontal circular path of radius $r = 2.00$ m with an angular speed, $\omega = 3.00$ rad/s. Her centripetal acceleration is

$$a_c = \frac{v_t^2}{r} = \frac{(r\omega)^2}{r} = r\omega^2$$

$$a_c = (2.00 \text{ m})(3.00 \text{ rad/s})^2 = 18.0 \text{ m/s}^2 \qquad \lozenge$$

(b)   The force, directed toward the center of the circular path, required to produce this centripetal acceleration is $F_c = ma_c$ or

$$F_c = (50.0 \text{ kg})(18.0 \text{ m/s}^2) = 900 \text{ N} \qquad \lozenge$$

(c)   Three forces act on the child as she stands on the rotating merry-go-round. These are: (1) a normal force, $\bar{n}$, exerted upward on the child by the merry-go-round; (2) a downward gravitational force

$$F_g = w = mg = (50.0 \text{ kg})(9.80 \text{ m/s}^2) = 490 \text{ N}$$

and (3) a horizontal friction force, $\vec{f}_s$, between her feet and the platform. Since the child's vertical acceleration is zero, Newton's second law gives

$$\sum F_y = n - F_g = 0 \qquad \text{or} \qquad n = F_g = 490 \text{ N}$$

The only force that is horizontal and thus capable of producing an acceleration toward the center of the horizontal circular path is the static friction force. Therefore, it is necessary that $f_s = F_c$. Since $f_s \leq \mu_s n$, this means that $F_c \leq \mu_s n$, and the required coefficient of static friction is

$$\mu_s \geq \frac{F_c}{n} = \frac{900 \text{ N}}{490 \text{ N}} = 1.84$$

Comparing this result to common values for μ_s (see Table 4.2 in the textbook), this required value for the coefficient of static friction is seen to be unrealistic. The child will not be able to stay on the merry-go-round. ◊

27.   A 40.0-kg child takes a ride on a Ferris wheel that rotates four times each minute and has a diameter of 18.0 m. (a) What is the centripetal acceleration of the child? (b) What force (magnitude and direction) does the seat exert on the child at the lowest point of the ride? (c) What force does the seat exert on the child at the highest point of the ride? (d) What force does the seat exert on the child when the child is halfway between the top and bottom?

Solution

(a)   The child moves in a circular path of radius $r = 9.00$ m with a constant angular velocity of $\omega = 4.00$ rev/min. The centripetal acceleration is

$$a_c = \frac{v_t^2}{r} = r\omega^2 = (9.00 \text{ m})\left[\left(4.00 \ \frac{\text{rev}}{\text{min}}\right)\left(\frac{2\pi \text{ rad}}{1 \text{ rev}}\right)\left(\frac{1 \text{ min}}{60 \text{ s}}\right)\right]^2 = 1.58 \text{ m/s}^2 \qquad \Diamond$$

(b)   At the lowest point on the ride, the forces acting on the child are a normal force exerted by the seat and the gravitational force. The resultant force must be directed toward the center of the path and have magnitude ma_c. Thus,

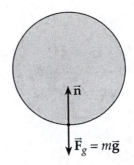

$$\Sigma F_y = n - mg = ma_c \qquad \text{or} \qquad n = m(g + a_c)$$

$$n = (40.0 \text{ kg})(9.80 \text{ m/s}^2 + 1.58 \text{ m/s}^2) = 455 \text{ N}$$

At this point on the ride, the seat exerts a force of 455 N directed upward toward the center of the Ferris wheel. $\Diamond$

(c)   At the highest point, the forces are as shown at the right. Again, the resultant force must be directed toward the center with magnitude ma_c. This yields

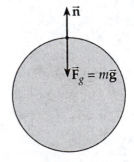

$$\Sigma F_y = n - mg = -ma_c \qquad \text{or} \qquad n = m(g - a_c)$$

$$n = (40.0 \text{ kg})(9.80 \text{ m/s}^2 - 1.58 \text{ m/s}^2) = 329 \text{ N}$$

At the top of the ride, the seat exerts a force of 329 N directed upward away from the center of the Ferris wheel. $\Diamond$

(d)   At a point halfway between the bottom and top, the seat must exert a force having an upward component $F_y = mg = 392\text{ N}$ and a horizontal component $F_x = ma_c = (40.0\text{ kg})(1.58\text{ m/s}^2) = 63.2\text{ N}$ as shown. Then, the resultant force acting on the child is again directed toward the center and has magnitude ma_c.

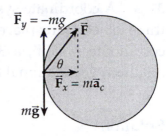

The magnitude of the resultant force exerted by the seat at this point is

$$F = \sqrt{F_x^2 + F_y^2} = \sqrt{(63.2\text{ N})^2 + (392\text{ N})^2} = 397\text{ N} \qquad \lozenge$$

and its direction is

$$\theta = \tan^{-1}\left(\frac{F_y}{F_x}\right) = \tan^{-1}\left(\frac{392\text{ N}}{63.2\text{ N}}\right) = 80.8° \text{ above the horizontal.} \qquad \lozenge$$

31.   A coordinate system (in meters) is constructed on the surface of a pool table, and three objects are placed on the table as follows: a 2.0-kg object at the origin of the coordinate system, a 3.0-kg object at (0, 2.0), and a 4.0-kg object at (4.0, 0). Find the resultant gravitational force exerted by the other two objects on the object at the origin.

Solution

Each of the other two objects will exert an attractive gravitation force on the object of mass $m_2 = 2.0$ kg located at the origin. Using Newton's law of universal gravitation, the magnitudes of these forces are found to be

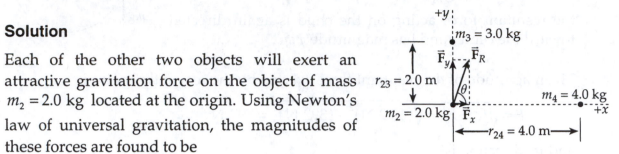

$$F_x = G\frac{m_2 m_4}{r_{24}^2} = \left(6.67\times10^{-11}\ \frac{\text{N}\cdot\text{m}^2}{\text{kg}^2}\right)\frac{(2.0\ \text{kg})(4.0\ \text{kg})}{(4.0\ \text{m})^2} = 3.3\times10^{-11}\ \text{N}$$

and

$$F_y = G\frac{m_2 m_3}{r_{23}^2} = \left(6.67\times10^{-11}\ \frac{\text{N}\cdot\text{m}^2}{\text{kg}^2}\right)\frac{(2.0\ \text{kg})(3.0\ \text{kg})}{(2.0\ \text{m})^2} = 1.0\times10^{-10}\ \text{N}$$

The resultant gravitational force exerted on m_2 by the other two objects is then

$$F_R = \sqrt{F_x^2 + F_y^2} = \sqrt{(3.3\times10^{-11}\ \text{N})^2 + (1.0\times10^{-10}\ \text{N})^2} = 1.1\times10^{-10}\ \text{N} \qquad \Diamond$$

directed at

$$\theta = \tan^{-1}\left(\frac{F_y}{F_x}\right) = \tan^{-1}\left(\frac{1.0\times10^{-10}\ \text{N}}{3.3\times10^{-11}\ \text{N}}\right) = 72° \text{ from the } +x \text{ direction.} \qquad \Diamond$$

35.   A satellite moves in a circular orbit around Earth at a speed of 5 000 m/s. Determine (a) the satellite's altitude above the surface of Earth and (b) the period of the satellite's orbit.

Solution

(a)   The gravitational force exerted on the satellite (mass m) by Earth (mass M_E) must produce the required centripetal acceleration, $a_c = v_t^2/r$, of the satellite. That is:

$$\frac{GM_E m}{r^2} = m\left(\frac{v_t^2}{r}\right) \qquad \text{which reduces to} \qquad r = \frac{GM_E}{v_t^2}$$

The radius of the satellite's orbit is therefore

$$r = \left(6.67\times10^{-11}\text{ N}\cdot\text{m}^2/\text{kg}^2\right)\frac{\left(5.98\times10^{24}\text{ kg}\right)}{\left(5\ 000\text{ m/s}\right)^2} = 1.60\times10^7\text{ m}$$

The altitude above Earth's surface, of radius R_E, is then

$$h = r - R_E = 1.60\times10^7\text{ m} - 6.38\times10^6\text{ m} = 9.58\times10^6\text{ m} \qquad \Diamond$$

(b)   The period is the time required for the satellite to complete one full trip around the circular orbit. This is:

$$T = \frac{\text{circumference of orbit}}{\text{orbital speed}} = \frac{2\pi r}{v_t}$$

$$T = \frac{2\pi\left(1.60\times10^7\text{ m}\right)}{5\ 000\text{ m/s}} = 2.00\times10^4\text{ s} = \left(2.00\times10^4\text{ s}\right)\left(\frac{1\text{ h}}{3600\text{ s}}\right) = 5.57\text{ h} \qquad \Diamond$$

39. A satellite of mass 200 kg is launched from a site on Earth's equator into an orbit 200 km above the surface of Earth. (a) Assuming a circular orbit, what is the orbital period of this satellite? (b) What is the satellite's speed in its orbit? (c) What is the minimum energy necessary to place the satellite in orbit, assuming no air friction?

Solution

The radius of the satellite's orbit is

$$r = R_E + 200 \text{ km} = 6.38 \times 10^6 \text{ m} + 200 \times 10^3 \text{ m} = 6.58 \times 10^6 \text{ m}$$

(a) The required centripetal acceleration of the satellite is produced by the gravitational force exerted on it by Earth.

Thus, $G \dfrac{M_E m}{r^2} = m \left(\dfrac{v_t^2}{r} \right)$ which yields $v_t = \sqrt{\dfrac{GM_E}{r}}$

The orbital speed of the satellite is therefore

$$v_t = \sqrt{\frac{\left(6.67 \times 10^{-11} \text{ N} \cdot \text{m}^2/\text{kg}^2\right)\left(5.98 \times 10^{24} \text{ kg}\right)}{6.58 \times 10^6 \text{ m}}} = 7.79 \times 10^3 \text{ m/s}$$

and its period is

$$T = \frac{\text{circumference}}{\text{speed}} = \frac{2\pi r}{v_t} = \frac{2\pi \left(6.58 \times 10^6 \text{ m}\right)}{7.79 \times 10^3 \text{ m/s}} = 5.31 \times 10^3 \text{ s} = 1.48 \text{ h} \qquad \Diamond$$

(b) The orbital speed of the satellite was found above to be

$$v_t = 7.79 \times 10^3 \text{ m/s} \qquad \Diamond$$

(c) When the satellite is in orbit, its gravitational potential energy is

$$\left(PE_g\right)_f = -G \frac{M_E m}{r} = -\left(6.67 \times 10^{-11} \frac{\text{N} \cdot \text{m}^2}{\text{kg}^2}\right) \frac{\left(5.98 \times 10^{24} \text{ kg}\right)\left(200 \text{ kg}\right)}{6.58 \times 10^6 \text{ m}}$$

$$= -1.21 \times 10^{10} \text{ J}$$

and its kinetic energy is

$$KE_f = \tfrac{1}{2} m v_t^2 = \tfrac{1}{2} \left(200 \text{ kg}\right)\left(7.79 \times 10^3 \text{ m/s}\right)^2 = 6.06 \times 10^9 \text{ J}$$

Initially the satellite was moving with the rotational speed of a point on the equator of Earth, or

$$v_i = \frac{2\pi R_E}{24.0 \text{ hr}} = \frac{2\pi \left(6.38 \times 10^6 \text{ m}\right)}{24.0 \left(3600 \text{ s}\right)} = 464 \text{ m/s}$$

Its initial kinetic energy was

$$KE_i = \tfrac{1}{2}mv_i^2 = \tfrac{1}{2}(200 \text{ kg})(464 \text{ m/s})^2 = 2.15 \times 10^7 \text{ J}$$

and its gravitational potential energy was $\left(PE_g\right)_i = -G\dfrac{M_E m}{R_e}$, or

$$\text{or} \quad \left(PE_g\right)_i = -\left(6.67 \times 10^{-11} \ \frac{\text{N} \cdot \text{m}^2}{\text{kg}^2}\right)\frac{\left(5.98 \times 10^{24} \text{ kg}\right)(200 \text{ kg})}{6.38 \times 10^6 \text{ m}} = -1.25 \times 10^{10} \text{ J}$$

The work-energy theorem then gives the energy required to place the satellite in orbit as $W_{nc} = (KE + PE)_f - (KE + PE)_i$

$$\text{or} \quad W_{nc} = \left(6.06 \times 10^9 \text{ J} - 1.21 \times 10^{10} \text{ J}\right) - \left(2.15 \times 10^7 \text{ J} - 1.25 \times 10^{10} \text{ J}\right)$$

$$W_{nc} = 6.43 \times 10^9 \text{ J}$$

$\Diamond$

47.   A car moves at speed v across a bridge made in the shape of a circular arc of radius r. (a) Find an expression for the normal force acting on the car when it is at the top of the arc. (b) At what minimum speed will the normal force become zero (causing the occupants of the car to seem weightless) if $r = 30.0$ m?

Solution

Consider the sketch at the right showing the forces acting on the car as it crosses the top of the arc of the bridge.

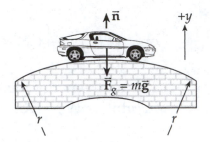

(a)   If the car remains in contact with the bridge, it follows a circular path of radius r. Thus, it must be accelerating **toward the center** of the circular path at a rate $a_c = v_t^2/r$.

When the car is at the top of the arc, the line toward the center is vertical. Taking upward as positive and applying Newton's second law to the vertical motion, $\sum F_y = ma_y$ yields

$$n - mg = m(-a_c) \qquad \text{or} \qquad n = m\left(g - \frac{v_t^2}{r}\right) \qquad \Diamond$$

(b)   If the occupants of the car (and the car itself) are to seem weightless, the downward force they exert on the roadway must be zero. Then, by Newton's third law, the upward force the roadway exerts on them (that is, the normal force n) must also be zero. Therefore, the desired speed is that for which

$$n = m\left(g - v_t^2/r\right) = 0 \quad \text{or} \quad v_t = \sqrt{rg}$$

If the radius of the circular arc is 30.0 m, then

$$v_t = \sqrt{(30.0 \text{ m})(9.80 \text{ m/s}^2)} = 17.1 \text{ m/s} \qquad \Diamond$$

51. In a popular amusement park ride, a rotating cylinder of radius 3.00 m is set in rotation at an angular speed of 5.00 rad/s, as in Figure P7.51. The floor then drops away, leaving the riders suspended against the wall in a vertical position. What minimum coefficient of friction between a rider's clothing and the wall is needed to keep the rider from slipping? (*Hint*: Recall that the magnitude of the maximum force of static friction is equal to $\mu_s n$, where n is the normal force—in this case, the force causing the centripetal acceleration.

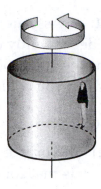

Figure P7.51

Solution

The normal force exerted on the person by the cylindrical wall must provide the centripetal acceleration, so

$$n = m\left(\frac{v_t^2}{r}\right) = \frac{m(r\omega)^2}{r} = m(r\omega^2)$$

If the minimum acceptable coefficient of friction is present, the person is on the verge of slipping and the maximum static friction force equals the person's weight, or

$$\left(f_s\right)_{max} = \left(\mu_s\right)_{min} n = mg$$

Thus,
$$\left(\mu_s\right)_{min} = \frac{mg}{n} = \frac{g}{r\omega^2} = \frac{9.80 \text{ m/s}^2}{(3.00 \text{ m})(5.00 \text{ rad/s})^2} = 0.131 \qquad \diamond$$

61. Assume that you are agile enough to run across a horizontal surface at 8.50 m/s, independently of the value of the gravitational field. What would be (a) the radius and (b) the mass of an airless spherical asteroid of uniform density 1.10×10^3 kg/m^3 on which you could launch yourself into orbit by running? (c) What would be your period?

Solution

(a) If you are to launch yourself into orbit by running, the minimum running speed is that for which the gravitational force acting on you exactly matches the force needed to produce the required centripetal acceleration. That is, we must have

$$G\frac{Mm}{R^2} = m\left(\frac{v_t^2}{R}\right)$$

where M is the mass of the asteroid and R is its radius. Since the mass of the asteroid may be written as

$$M = \text{density} \times \text{volume} = \rho\left(\frac{4}{3}\pi R^3\right)$$

this requirement becomes

$$G\rho\left(\frac{4}{3}\pi R^3\right)\frac{\cancel{m}}{R^2} = \cancel{m}\left(\frac{v_t^2}{R}\right) \qquad \text{or} \qquad R = \sqrt{\frac{3v_t^2}{4\pi G\rho}}$$

The radius of the asteroid would then be

$$R = \sqrt{\frac{3(8.50 \text{ m/s})^2}{4\pi\left(6.67\times10^{-11} \text{ N}\cdot\text{m}^2/\text{kg}^2\right)\left(1.10\times10^3 \text{ kg/m}^3\right)}}$$

or $\qquad R = 1.53\times10^4$ m $= 15.3$ km $\qquad\qquad\qquad\qquad\qquad\qquad\qquad\qquad\qquad$ ◊

(b) The mass of the asteroid will be

$$M = \rho\left(\frac{4}{3}\pi R^3\right) = \frac{4}{3}\left(1.10\times10^3 \text{ kg/m}^3\right)\pi\left(1.53\times10^4 \text{ m}\right)^3 = 1.66\times10^{16} \text{ kg} \qquad ◊$$

(c) Your orbital period would be

$$T = \frac{2\pi}{\omega} = \frac{2\pi R}{v_t} = \frac{2\pi\left(1.53\times10^4 \text{ m}\right)}{8.50 \text{ m/s}} = 1.13\times10^4 \text{ s} = 3.15 \text{ h} \qquad ◊$$

Chapter 8

Rotational Equilibrium and Rotational Dynamics

NOTES FROM SELECTED CHAPTER SECTIONS

8.1 Torque

Torque is the physical quantity which is a measure of the tendency of a force to cause rotation of a body about a specified axis. **Torque must be defined with respect to a specific axis of rotation**. Torque, which has the SI **units** of N·m, must not be confused with work and energy.

8.2 Torque and the Two Conditions For Equilibrium

A body in static equilibrium must satisfy two conditions:

1. The resultant external force must be zero.

2. The resultant external torque about any axis must be zero.

8.3 The Center of Gravity

In order to calculate the torque due to the weight (gravitational force) on a rigid body, the entire weight of the object can be considered to be concentrated at a single point called the center of gravity. *The center of gravity of a homogeneous, symmetric body must lie along an axis of symmetry*

8.5 Relationship Between Torque and Angular Acceleration

The angular acceleration of an object is proportional to the net torque acting on it. The moment of inertia of the object is the proportionality constant between the net torque and the angular acceleration. **The force and mass in linear motion correspond to torque and moment of inertia in rotational motion**. Moment of inertia of an object depends on the location of the axis of rotation and upon the manner in which the mass is distributed relative to that axis (e.g., a ring has a greater moment of inertia than a disk of the same mass and radius).

8.6   Rotational Kinetic Energy

In linear motion, the energy concept is useful in describing the motion of a system. The energy concept can be equally useful in simplifying the analysis of rotational motion. We now have expressions for four types of mechanical energy: gravitational potential energy, PE_g; elastic potential energy, PE_s; translational kinetic energy, KE_t; and rotational kinetic energy, KE_r. We must include all these forms of energy in the equation for conservation of mechanical energy.

EQUATIONS AND CONCEPTS

The **magnitude of the torque** about a specified axis due to a given force depends on:

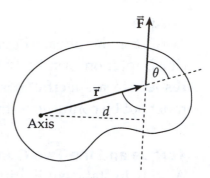

- magnitude of the force
- point of application of the force
- direction of the force

The **lever arm** is the perpendicular distance from the axis of rotation to the line along which the force is acting.

$$\tau = Fr\sin\theta \qquad (8.2)$$

See figure at right.

$$\text{lever arm} = d = r\sin\theta$$

The **algebraic sign of a torque** due to a force is considered positive if the force has a tendency to rotate the body counterclockwise about the chosen axis, and negative if the tendency for rotation is clockwise.

Sign conventions for torque

The **first condition for equilibrium** requires that the net external force acting on a body equal zero. This will ensure that the body is in translational equilibrium.

$$\sum \vec{F} = 0$$

The **second condition for equilibrium** requires that the net external torque acting on a body equals zero. This condition will ensure that the body is in rotational equilibrium.

$$\sum \vec{\tau} = 0$$

To compute the torque due to the force of gravity (an object's weight), the total weight can be considered as being concentrated at a single point called the **center of gravity** and having coordinates x_{cg}, y_{cg}, z_{cg}. *The center of gravity of a symmetric homogeneous body must lie on the axis of symmetry.*

$$x_{cg} = \frac{\sum m_i x_i}{\sum m_i} \qquad (8.3a)$$

$$y_{cg} = \frac{\sum m_i y_i}{\sum m_i} \qquad (8.3b)$$

$$z_{cg} = \frac{\sum m_i z_i}{m_i} \qquad (8.3c)$$

The **angular acceleration** of a point mass, moving in a path of radius r, is proportional to the net torque acting on the mass.

$$\tau = mr^2 \alpha \qquad (8.6)$$

For a **point mass** m, at a radial distance r, from a specified axis of rotation, $I = mr^2$.

For a **collection of discrete point masses**, each with its own corresponding values of m and r, $I = \sum\limits_{i} m_i r_i^2$.

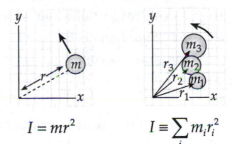

$$I = mr^2 \qquad\qquad I \equiv \sum\limits_{i} m_i r_i^2$$

The **moment of inertia of a rigid body** depends on the mass and the distribution of mass relative to the axis of rotation.

$$I \equiv \sum mr^2 \qquad (8.7)$$

The value of I can be calculated for an extended object with good symmetry. Expressions for the moments of inertia for a number of objects of common shape are given in Table 8.1 of your text. The SI units of moment of inertia are $kg \cdot m^2$.

The **angular acceleration** of an extended object is proportional to the net torque acting on the object.

$$\sum \tau = I\alpha \qquad (8.8)$$

A rigid body or mass in rotational motion has **rotational kinetic energy** due to its motion. *Note that this is not a new form of energy but is a convenient form for representing kinetic energy associated with rotational motion.*

$$KE_r \equiv \tfrac{1}{2}I\omega^2 \tag{8.10}$$

When only conservative forces act on a system, the total mechanical energy of the system (gravitational potential, elastic potential, rotational kinetic and translational kinetic) is conserved.

$$(KE_t + KE_r + PE)_i = \tag{8.11}$$
$$(KE_t + KE_r + PE)_f$$

Angular momentum, $\vec{L}$ is a vector quantity associated with rotational motion. *Angular momentum has the direction of the angular velocity.*

$$L \equiv I\omega \tag{8.13}$$

The net external torque acting on an object equals the time rate of change of its angular momentum. *Equation 8.14 is the rotational analog of Newton's second law for transitional motion.*

$$\sum \tau = \frac{\Delta L}{\Delta t} \tag{8.14}$$

The **angular momentum is conserved** when the net external torque acting on a system is zero.

$$L_i = L_f \qquad \text{if} \qquad \sum \tau = 0 \tag{8.15}$$
$$I_i\omega_i = I_f\omega_f \qquad \text{if} \qquad \sum \tau = 0 \tag{8.16}$$

SUGGESTIONS, SKILLS, AND STRATEGIES

PROBLEM-SOLVING STRATEGY FOR OBJECTS IN EQUILIBRIUM

1. Draw a simple, neat diagram of the system that is large enough to show all the forces clearly.

2. Isolate the object of interest being analyzed. Draw a free-body diagram for this object showing all external forces acting on the object. For systems containing more than one object, draw separate diagrams for each object. Do not include forces that the object exerts on its surroundings.

3. Establish convenient coordinate axes for each body and find the components of the forces along these axes. Now apply the first condition of equilibrium for each object under consideration; namely, that the net force on the object in the x and y directions must be zero.

4. Choose a convenient origin (axis of rotation) for calculating the net torque on the object. Now apply the second condition of equilibrium that says that the net torque on the object about any origin must be zero. Remember that the choice of the origin for the torque equation is arbitrary; therefore, choose an origin that will simplify your calculation as much as possible. Note that a force that acts along a line passing through the point chosen as the axis of rotation gives zero contribution to the torque because the lever arm is zero in this case.

5. The first and second conditions for equilibrium will give a set of simultaneous equations with several unknowns. To complete your solution, all that is left is to solve for the unknowns in terms of the known quantities.

PROBLEM-SOLVING STRATEGY FOR ROTATIONAL MOTION

The following facts and procedures should be kept in mind when solving rotational motion problems.

1. Problems involving the equation $\Sigma \vec{\tau} = I\vec{\alpha}$ are very similar to those encountered in Newton's second law problems, $\Sigma \vec{F} = m\vec{a}$. Note the correspondences between linear and rotational quantities in that $\vec{F}$ is replaced by $\vec{\tau}$ m by I, and $\vec{a}$ by $\vec{\alpha}$.

2.   Other analogues between rotational quantities and linear quantities include the replacement of x by θ and v by ω. Recall that each linear quantity (x, v, and a) equals the product of the radius and the corresponding angular quantity (θ, ω, and α). These are helpful as memory devices for such rotational motion quantities as rotational kinetic energy, $KE_r = \frac{1}{2}I\omega^2$, and angular momentum, $L = I\omega$.

3.   With the analogues mentioned in Step 2, conservation of energy techniques remain the same as those examined in Chapter 5, except for the fact that a new kind of energy, rotational kinetic energy, must be included in the expression for the conservation of energy (see Equation 8.11).

4.   Likewise, the techniques for solving conservation of angular momentum problems are essentially the same as those used in solving conservation of linear momentum problems, except you are equating total angular momentum before to total angular momentum after as $I_i\omega_i = I_f\omega_f$.

REVIEW CHECKLIST

▷   There are two necessary conditions for equilibrium of a rigid body: $\Sigma F = 0$ and $\Sigma \tau = 0$. Torques which cause counterclockwise rotations are positive and those causing clockwise rotations are negative.

▷   The torque associated with a force has a magnitude equal to the force times the lever arm. The lever arm is the perpendicular distance from the axis of rotation to a line drawn along the direction of the force. Also, the net torque on a rigid body about some axis is proportional to the angular acceleration; that is, $\Sigma \tau = I\alpha$, where I is the moment of inertia about the axis for which the net torque is evaluated.

▷   The work-energy theorem can be applied to a rotating rigid body. That is, the net work done on a rigid body rotating about a fixed axis equals the change in its rotational kinetic energy. The law of conservation of mechanical energy can be used in the solution of problems involving rotating rigid bodies.

▷   The time rate of change of the angular momentum of a rigid body rotating about an axis is proportional to the net torque acting about the axis of rotation. This is the rotational analog of Newton's second law.

SOLUTIONS TO SELECTED END-OF-CHAPTER PROBLEMS

3. Calculate the net torque (magnitude and direction) on the beam in Figure P8.3 about (a) an axis through O perpendicular to the page and (b) an axis through C perpendicular to the page.

Solution

To solve this problem in the easiest manner, first resolve all of the forces shown in Figure P8.3 into components parallel to and perpendicular to the beam as shown below. Then, it is observed that most of these components have zero lever arms about an axis perpendicular to the page through point O; and the lever arms that are not zero are easily determined. This is also true for a similar axis through point C.

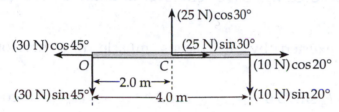

(a) $\tau_O = +\left[(25\text{ N})\cos 30°\right](2.0\text{ m}) - \left[(10\text{ N})\sin 20°\right](4.0\text{ m}) = +30\text{ N}\cdot\text{m}$

  or $\tau_O = 30\text{ N}\cdot\text{m}$ counterclockwise $\Diamond$

(b) $\tau_C = +\left[(30\text{ N})\sin 45°\right](2.0\text{ m}) - \left[(10\text{ N})\sin 20°\right](2.0\text{ m}) = +36\text{ N}\cdot\text{m}$

  or $\tau_C = 36\text{ N}\cdot\text{m}$ counterclockwise $\Diamond$

9. A cook holds a 2.00-kg carton of milk at arm's length (Fig. P8.9). What force $\vec{F}_B$ must be exerted by the biceps muscle? (Ignore the weight of the forearm.)

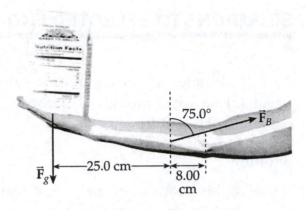

Figure P8.9

Solution

For the system consisting of the forearm and the milk carton to be in equilibrium, it is necessary that the sum of the torques be zero about any axis we choose.

The upper arm exerts some unknown force on the forearm at the elbow. However, if we select a rotation axis that is perpendicular to the page and passes through the elbow, this unknown force will have zero lever arm and exert zero torque about our axis.

Resolve the force exerted by the biceps muscle into horizontal and vertical components:

$$(F_B)_x = F_B \sin 75.0° \qquad \text{and} \qquad (F_B)_y = F_B \cos 75.0°$$

The line along which the horizontal component acts passes through the chosen rotation axis, so that component exerts zero torque. Applying the second condition for equilibrium to the system (forearm and milk carton) yields

$$\sum \tau = + F_g (25.0 \text{ cm} + 8.00 \text{ cm}) - (F_B)_y (8.00 \text{ cm}) = 0$$

or $\quad mg(25.0 \text{ cm} + 8.00 \text{ cm}) - (F_B \cos 75.0°)(8.00 \text{ cm}) = 0$

The force exerted by the biceps muscle is then given by

$$F_B = \frac{\left[(2.00 \text{ kg})(9.80 \text{ m/s}^2)\right](33.0 \text{ cm})}{(8.00 \text{ cm})\cos 75.0°} = 312 \text{ N}$$

◊

17.   A 500-N uniform rectangular sign 4.00 m wide and 3.00 m high is suspended from a horizontal, 6.00-m-long, uniform, 100 N rod as indicated in Figure P8.17. The left end the rod is supported by a hinge, and the right end is supported by a thin cable making a 30.0° angle with the vertical. (a) Find the tension T in the cable. (b) Find the horizontal and vertical components of force exerted on the left end of the rod by the hinge.

Solution

The free-body diagram of the sign-rod combination is given below. Note that the tension in the cable and the reaction force exerted on the left end of the rod by the hinge have been resolved into horizontal and vertical components.

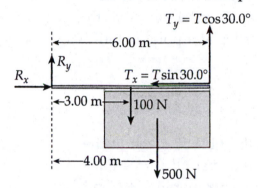

Consider a rotation axis perpendicular to the page and passing through the left end of the rod and apply the second condition of equilibrium, $\Sigma\tau = 0$, to this system. This gives

$$+(T\cos 30.0°)(6.00\ \text{m}) - (100\ \text{N})(3.00\ \text{m}) - (500\ \text{N})(4.00\ \text{m}) = 0$$

or $T = \dfrac{2.30 \times 10^3\ \text{N} \cdot \text{m}}{(6.00\ \text{m})\cos 30.0°} = 443\ \text{N}$ ◊

Now, apply the first condition of equilibrium to this system:

$$\Sigma F_x = 0 \implies R_x - T\sin 30.0° = 0$$

or $R_x = +(443\ \text{N})\sin 30.0° = +222\ \text{N} = 222\ \text{N}$ toward the right   ◊

and $\Sigma F_y = 0 \implies R_y + T\cos 30.0° - 100\ \text{N} - 500\ \text{N} = 0$

or $R_y = 600\ \text{N} - (443\ \text{N})\cos 30.0° = +216\ \text{N} = 216\ \text{N}$ upward   ◊

27.   The large quadriceps muscle in the upper leg terminates at its lower end in a tendon attached to the upper end of the tibia (Fig. P8.27a). The forces on the lower leg when the leg is extended are modeled as in Figure P8.27b, where $\vec{T}$ is the force of tension in the tendon, $\vec{w}$ is the force of gravity acting on the lower leg, and $\vec{F}$ is the force of gravity acting on the foot. Find $\vec{T}$ when the tendon is at an angle of 25.0° with the tibia, assuming that $w = 30.0$ N, $F = 12.5$ N, and the leg is extended at an angle θ of 40.0° with the vertical. Assume that the center of gravity of the lower leg is at its center and that the tendon attaches to the lower leg at a point one-fifth of the way down the leg.

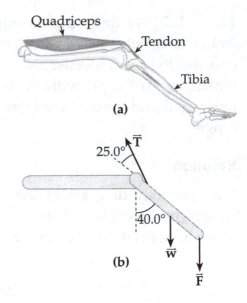

(a)

(b)

Figure P8.27

Solution

The free-body diagram of the tibia is given at the right. Note that $\theta = 40.0°$ and that all forces have been resolved into components parallel to and perpendicular to the tibia. We shall choose an axis that is perpendicular to the page and passing through the upper end of the tibia. Only T_y, w_y, and F_y have non-zero torques about this axis. The magnitudes of these components are:

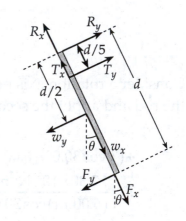

$$T_y = T\sin 25.0°$$

$$w_y = w\sin 40.0° = (30.0 \text{ N})\sin 40.0° = 19.3 \text{ N}$$

and $F_y = F\sin 40.0° = (12.5 \text{ N})\sin 40.0° = 8.03 \text{ N}$

When the tibia is held in rotational equilibrium in the position shown, the second condition of equilibrium ($\Sigma\tau = 0$) gives the tension in the tendon as:

$$+(T\sin 25.0°)\frac{d}{5} - (19.3 \text{ N})\frac{d}{2} - (8.03 \text{ N})d = 0$$

or $T = 5\left(\dfrac{17.7 \text{ N}}{\sin 25.0°}\right) = 209$ N   ◊

35. A 150-kg merry-go-round in the shape of a uniform, solid, horizontal disk of radius 1.50 m is set in motion by wrapping a rope about the rim of the disk and pulling on the rope. What constant force must be exerted on the rope to bring the merry-go-round from rest to an angular speed of 0.500 rev/s in 2.00 s?

Solution

The sketch at the right gives a view of the merry-go-round from above. The moment of inertia of a uniform, solid disk of mass $M = 150$ kg an radius $r = 1.50$ m is:

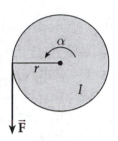

$$I = \frac{1}{2}Mr^2 = \frac{1}{2}(150 \text{ kg})(1.5 \text{ m})^2 = 169 \text{ kg} \cdot \text{m}^2$$

The desired angular acceleration for the merry-go-round is:

$$\alpha = \frac{\omega - \omega_0}{\Delta t} = \frac{0.500 \text{ rev/s} - 0}{2.00 \text{ s}} = 0.250 \frac{\text{rev}}{\text{s}^2}\left(\frac{2\pi \text{ rad}}{1 \text{ rev}}\right) = 1.57 \text{ rad/s}^2$$

We take counterclockwise torques (and hence, angular accelerations) as positive, and use the rotational form of Newton's second law to find the tension in the rope required to produce this angular acceleration:

$$\Sigma\tau = Fr = I\alpha \quad \text{or} \quad F = \frac{I\alpha}{r} = \frac{(169 \text{ kg} \cdot \text{m})^2(1.57 \text{ rad/s}^2)}{1.50 \text{ m}} = 177 \text{ N} \qquad \Diamond$$

39.   A 10.0-kg cylinder rolls without slipping on a rough surface. At the instant when its center of gravity has a speed of 10.0 m/s, determine (a) the translational kinetic energy of its center of gravity, (b) the rotational kinetic energy about its center of gravity, and (c) its total kinetic energy.

Solution

(a)   The translational kinetic energy of the cylinder is given by $KE_t = \frac{1}{2}mv^2$, where v is the translational speed of the center of gravity. Thus,

$$KE_t = \frac{1}{2}(10.0 \text{ kg})(10.0 \text{ m/s})^2 \qquad \text{or} \qquad KE_t = 500 \text{ J} \qquad \lozenge$$

(b)   The rotational kinetic energy of the cylinder is $KE_r = \frac{1}{2}I\omega^2$ where I is the moment of inertia about the axis through its center of gravity, and ω is the angular speed about this axis. Assuming a uniform, solid cylinder of radius R,

$$I = \frac{1}{2}mR^2$$

and the rotational kinetic energy is $KE_r = \frac{1}{2}\left(\frac{1}{2}mR^2\right)\omega^2 = \frac{1}{4}m(R\omega)^2$

The product $R\omega$ is the same as the tangential speed of a point on the rim of the cylinder, $v_t = R\omega$, so $KE_r = \frac{1}{4}mv_t^2$

If the wheel rolls without slipping, the tangential speed of a point on the rim is the same as the translational speed of the center of gravity. To understand why this is true, imagine yourself to be at the center of the wheel and moving to your left at speed v. Looking down, you would see point A at the wheel's rim moving to your right with the tangential speed $v_t = R\omega$. You would also see point B on the ground moving to your right at the speed v, the speed of the center of gravity (and you) relative to the ground.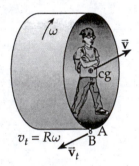

Now, if the rim of the wheel is not slipping against the ground, the points A and B (in contact with each other) must move at the same speed. Thus, it is necessary that $v = v_t = R\omega$ if slipping does not occur. The rotational kinetic energy of the wheel is therefore,

$$KE_r = \frac{1}{4}mv_t^2 = \frac{1}{4}mv^2 = \frac{1}{4}(10.0 \text{ kg})(10.0 \text{ m/s})^2 = 250 \text{ J} \qquad \lozenge$$

(c)   The total kinetic energy of the rolling wheel is then:

$$KE_{\text{total}} = KE_t + KE_r = 500 \text{ J} + 250 \text{ J} = 750 \text{ J} \qquad \lozenge$$

43.   The top in Figure P8.43 has a moment of inertia of 4.00×10^{-4} kg·m^2 and is initially at rest. It is free to rotate about a stationary axis, AA'. A string wrapped around a peg along the axis of the top is pulled in such a manner as to maintain a constant tension of 5.57 N in the string. If the string does not slip while wound around the peg, what is the angular speed of the top after 80.0 cm of string has been pulled off the peg? [**Hint:** Consider the work done.]

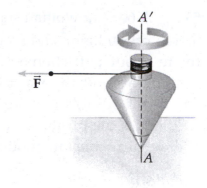

Figure P8.43

Solution

The agent pulling on the end of the string does work on the system consisting of the string and top. When the end of the string has been moved distance s, the work done on this system is $W_{net} = Fs\cos 0°$, where F is the constant tension maintained in the string.

Neglecting the mass of the string, the only kinetic energy of the system is the rotational kinetic energy of the top. The work-energy theorem, $W_{net} = KE_f - KE_i$, then becomes

$$Fs\cos 0° = \tfrac{1}{2}I_{top}\omega_f^{\,2} - \tfrac{1}{2}I_{top}\omega_i^2$$

If the top starts from rest and the tension maintained in the string is $F = 5.57$ N, the angular speed of the top after the end of the string has been moved 80.0 cm is

$$\omega_f = \sqrt{\frac{2Fs\cos 0°}{I_{top}}} = \sqrt{\frac{2(5.57 \text{ N})(0.800 \text{ m})(1)}{4.00 \times 10^{-4} \text{ kg·m}^2}} = 149 \text{ rad/s} \qquad \Diamond$$

53.   A 60.0-kg woman stands at the rim of a horizontal turntable having a moment of inertia of 500 kg·m^2 and a radius of 2.00 m. The turntable is initially at rest and is free to rotate about a frictionless, vertical axle through its center. The woman then starts walking around the rim clockwise (as viewed from above the system) at a constant speed of 1.50 m/s relative to the Earth. (a) In what direction and with what angular speed does the turntable rotate? (b) How much work does the woman do to set herself and the turntable into motion?

Solution

(a)   If no external agent exerts a torque about the vertical axis through the center of the turntable, the total angular momentum of the system (woman plus turntable) about this axis remains constant. When the woman is at rest on the stationary turntable, the total angular momentum of the system is zero $(L_i = 0)$. As the woman starts walking around the axis, the turntable must develop a counterclockwise angular momentum whose magnitude equals the magnitude of the woman's clockwise angular momentum. These two contributions to the total angular momentum will then cancel each other.

Treating the woman as a point object on the rim of the turntable (at distance $r = 2.00$ m from the axis), her moment of inertia about the central axis is $I_w = m_w r^2$. Taking counterclockwise angular momentum as positive, the woman's clockwise angular momentum as she walks at constant speed v relative to Earth is

$$L_w = -I_w \omega_w = -\left(m_w r^2\right)\left(\frac{v}{r}\right) = -m_w rv$$

The angular momentum of the turntable is $L_t = I_t \omega_t$ and conservation of angular momentum requires that

$$L_{final} = \left(L_t + L_w\right) = L_i = 0 \qquad \text{or} \qquad I_t \omega_t - m_w rv = 0$$

Thus $\omega_t = \dfrac{m_w rv}{I_t} = \dfrac{(60.0 \text{ kg})(2.00 \text{ m})(1.50 \text{ m/s})}{500 \text{ kg·m}^2} = +0.36$ rad/s

or $\omega_t = 0.36$ rad/s counterclockwise   ◊

(b)   The work-energy theorem gives the work done by the woman to set this system in motion as

$$W_{net} = KE_f - KE_i = \left(\tfrac{1}{2}I_t \omega_t^2 + \tfrac{1}{2}m_w v^2\right) - 0$$

$$W_{net} = \frac{1}{2}\left(500 \text{ kg·m}^2\right)(0.36 \text{ rad/s})^2 + \frac{1}{2}(60.0 \text{ kg})(1.50 \text{ m/s})^2 = 99.9 \text{ J} \qquad ◊$$

63. A solid 2.0-kg ball of radius 0.50 m starts at a height of 3.0 m above the surface of the Earth and **rolls** down a 20° slope. A solid disk and a ring start at the same time and the same height. The ring and disk each have the same mass and radius as the ball. Which of the three wins the race to the bottom if all roll without slipping?

Solution

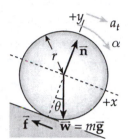

Consider the free-body diagram of an object rolling down the incline. The object could be either a solid sphere (ball), a solid cylinder (disk), or a hoop (ring). If the object rolls without slipping, then $a_t = r\alpha$ where a_t is the linear acceleration of the center of gravity and α is the angular acceleration about the rotation axis.

From $\Sigma F_x = ma_x$ we obtain
$$mg\sin\theta - f = ma_t \qquad \text{[1]}$$

Now, consider an axis perpendicular to the page and passing through the center of the object.

$\Sigma\tau = I\alpha$ becomes
$$f \cdot r = I\alpha = I\left(\frac{a_t}{r}\right) \quad \text{or} \quad f = \left(\frac{I}{r^2}\right)a_t$$

Substitute this result into Equation [1] and simplify to obtain
$$a_t = \frac{g\sin\theta}{\left(1 + I/mr^2\right)}$$

as the linear acceleration of the center of gravity of the object.

For a solid sphere,
$$I = \tfrac{2}{5}mr^2 \qquad \text{so} \qquad a_{\text{sphere}} = \frac{g\sin\theta}{1.4}$$

For a solid cylinder,
$$I = \tfrac{1}{2}mr^2 \qquad \text{so} \qquad a_{\text{cylinder}} = \frac{g\sin\theta}{1.5}$$

Finally, for a hoop,
$$I = mr^2 \qquad \text{so} \qquad a_{\text{ring}} = \frac{g\sin\theta}{2.0}$$

Thus, we find $a_{\text{sphere}} > a_{\text{cylinder}} > a_{\text{ring}}$, so the sphere wins the race, the disk comes in second, and the ring is last. ◊

Note that each of the calculated accelerations is independent of the both the mass and the radius of the object. Thus, the three objects need not be of the same mass or size. Only the shapes and the distribution of the mass within the objects affect the outcome of the race.

66.   (a) Without the wheels, a bicycle frame has a mass of 8.44 kg. Each of the wheels can be roughly modeled as a uniform solid disk with a mass of 0.820 kg and a radius of 0.343 m. Find the kinetic energy of the whole bicycle when it is moving forward at 3.35 m/s. (b) Before the invention of a wheel turning on an axle, ancient people moved heavy loads by placing rollers under them. (Modern people use rollers, too: Any hardware store will sell you a roller bearing for a lazy Susan.) A stone block of mass 844 kg moves forward at 0.335 m/s, supported by two uniform cylindrical tree trunks, each of mass 82.0 kg and radius 0.343 m. There is no slipping between the block and the rollers or between the rollers and the ground. Find the total kinetic energy of the moving objects.

Solution

(a)   The frame and the center of each wheel move forward at $v = 3.35$ m/s and each wheel also turns at angular speed $\omega = v/R$. The total kinetic energy of the bicycle is $KE = KE_t + KE_r$, or

$$KE = \frac{1}{2}\left(m_{frame} + 2m_{wheel}\right)v^2 + 2\left(\frac{1}{2}I_{wheel}\omega^2\right)$$

$$= \frac{1}{2}\left(m_{frame} + 2m_{wheel}\right)v^2 + \frac{1}{2}\left(m_{wheel}R^2\right)\left(\frac{v^2}{R^2}\right)$$

This yields

$$KE = \frac{1}{2}\left(m_{frame} + 3m_{wheel}\right)v^2 = \frac{1}{2}\left[8.44\text{ kg} + 3(0.820\text{ kg})\right](3.35\text{m/s})^2 = 61.2\text{ J}$$ ◊

(b)   Since the block does not slip on the roller, its forward speed must equal that of point A, the uppermost point on the rim of the roller. That is, $v = |\vec{v}_{AE}|$ where $\vec{v}_{AE}$ is the velocity of A relative to Earth.

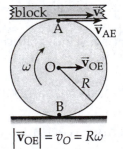

Please refer to the discussion in Part (b) of Problem 39 earlier in this chapter. Since the roller does not slip on the ground, the velocity of point O (the roller center) must have the same magnitude as the tangential speed of point B (the point on the roller rim in contact with the ground). That is, $|\vec{v}_{OE}| = R\omega = v_O$. Also, note that the velocity of point A relative to the roller center has a magnitude equal to the tangential speed $R\omega$, or $|\vec{v}_{AO}| = R\omega = v_O$.

From the discussion of relative velocities in Chapter 3, we know that $\vec{v}_{AE} = \vec{v}_{AO} + \vec{v}_{OE}$. Since all of these velocities are in the same direction, we may add their magnitudes getting $|\vec{v}_{AE}| = |\vec{v}_{AO}| + |\vec{v}_{OE}|$, or $v = v_O + v_O = 2v_O = 2R\omega$. Thus, we have determined that the translational speed of the center of the rollers (trees) is one half the speed of the stone $(v_O = v/2)$. Then, the angular speed of the roller is $\omega = v_O/R = v/2R$.

The total kinetic energy is $KE = KE_{translation} + KE_{rotation}$, or

$$KE = \frac{1}{2}m_{stone}v^2 + 2\left[\frac{1}{2}m_{tree}\left(\frac{v}{2}\right)^2\right] + 2\left(\frac{1}{2}I_{tree}\omega^2\right)$$

$$= \left(\frac{1}{2}m_{stone} + \frac{1}{4}m_{tree}\right)v^2 + \frac{1}{2}m_{tree}R^2\left(\frac{v^2}{4R^2}\right) = \frac{1}{2}\left(m_{stone} + \frac{3}{4}m_{tree}\right)v^2$$

This gives $KE = \frac{1}{2}\left[844\text{ kg} + \frac{3}{4}(82.0\text{ kg})\right](0.335\text{ m/s})^2 = 50.8\text{ J}$ ◊

73. A uniform solid cylinder of mass M and radius R rotates on a frictionless horizontal axle (Fig. P8.73). Two objects with equal masses hang from light cords wrapped around the cylinder. If the system is released from rest, find (a) the tension in each cord and (b) the acceleration of each object after the objects have descended a distance of h.

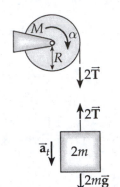

Figure P8.73

Solution

In the free-body diagrams at the right, the two falling masses have been combined into one object of mass $2m$. The total downward force these masses exert on the cylinder is $2T$ where T is the tension in either cord. If the cords do not slip on the cylinder, the linear acceleration of each falling object is related to the angular acceleration of the cylinder by $a_t = R\alpha$.

Choose an axis perpendicular to the page and passing through the center of the cylinder. Then, applying $\Sigma\tau = I\alpha$ to the cylinder gives

$$(2T)R = \left(\tfrac{1}{2}MR^2\right)\alpha = \left(\tfrac{1}{2}MR^2\right)\left(\frac{a_t}{R}\right) \qquad \text{or} \qquad T = \tfrac{1}{4}Ma_t \qquad\qquad [1]$$

Now apply $\Sigma F_y = ma_y$ to the falling objects to obtain

$$(2m)g - 2T = (2m)a_t \qquad \text{or} \qquad a_t = g - \frac{T}{m} \qquad\qquad [2]$$

(a) Substituting Equation [2] into Equation [1] yields

$$T = \frac{Mg}{4} - \left(\frac{M}{4m}\right)T \qquad \text{which reduces to} \qquad T = \frac{Mmg}{M+4m} \qquad\qquad \Diamond$$

(b) From Equation [2], the acceleration of each falling object is found to be

$$a_t = g - \frac{1}{m}\left(\frac{Mmg}{M+4m}\right) = g - \frac{Mg}{M+4m} = \frac{4mg}{M+4m} \qquad\qquad \Diamond$$

Note that the results are independent of the distance fallen, h.

80. A string is wrapped around a uniform cylinder of mass M and radius R. The cylinder is released from rest with the string vertical and its top end tied to a fixed bar (Fig. P8.80). Show that (a) the tension in the string is one-third the weight of the cylinder, (b) the magnitude of the acceleration of the center of gravity is $2g/3$, and (c) the speed of the center of gravity is $(4gh/3)^{1/2}$ after the cylinder has descended through distance h. Verify your answer to (c) with the energy approach.

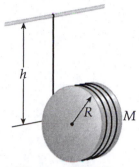

Figure P8.80

Solution

(a) The cylinder will rotate about an axis perpendicular to the page and through its center of gravity as the center of gravity accelerates downward. Applying Newton's second law to both the translational motion and the rotational motion gives:

From $\Sigma F_y = ma_y$, $\qquad T - Mg = M(-a)$

or $\qquad T = M(g-a)$ [1]

From $\Sigma\tau = I\alpha$, $\qquad -TR = \left(\dfrac{1}{2}MR^2\right)\left(-\dfrac{a}{R}\right)$

or $\qquad a = \dfrac{2T}{M}$ [2]

Substituting Equation [2] into Equation [1] and solving for the tension yields

$\qquad T = Mg/3$ ◊

(b) From Equation [2], $\qquad a = \dfrac{2T}{M} = \dfrac{2(Mg/3)}{M} = 2g/3$ ◊

(c) From $v_y^2 = v_{0y}^2 + 2a_y\Delta y$, $\qquad v = \sqrt{0 + 2\left(-\dfrac{2g}{3}\right)(-h)} = \sqrt{\dfrac{4gh}{3}}$ ◊

Using the work-energy theorem, $W_{net} = KE_f - KE_i$, we have

$\qquad Mgh - Th = \dfrac{1}{2}Mv^2 - 0$

or $\qquad v = \sqrt{\dfrac{2Mgh - 2(Mg/3)h}{M}} = \sqrt{\dfrac{4gh}{3}}$ ◊

Chapter 9

Solids and Fluids

NOTES FROM SELECTED CHAPTER SECTIONS

9.1 States of Matter

Matter is generally classified as being in one of three states: solid, liquid, or gas.

In a **crystalline solid,** the atoms are arranged in an ordered periodic structure; while in an **amorphous solid** (i.e. glass), the atoms are present in a disordered fashion.

In the **liquid state,** thermal agitation is greater than in the solid state, the molecular forces are weaker, and molecules wander throughout the liquid in a random fashion.

The molecules of a **gas** are in constant random motion and exert weak forces on each other. The distances separating molecules are large compared to the dimensions of the molecules.

9.2 The Deformation of Solids

The **elastic properties of solids** are described in terms of stress and strain. Stress is a quantity that is related to the force causing a deformation; strain is a measure of the degree of deformation. It is found that, for sufficiently small stresses, stress is proportional to strain, and the constant of proportionality depends on the material being deformed and on the nature of the deformation. We call this proportionality constant the elastic modulus.

We shall consider three types of deformation and define an elastic modulus for each:

- **Young's modulus,** which measures the resistance of a solid to a change in its length.

- **Shear modulus,** which measures the resistance to displacement of the planes of a solid sliding past each other.

- **Bulk modulus,** which measures the resistance that solids or liquids offer to changes in their volume.

9.3   Density and Pressure

The **density,** ρ, of a substance of uniform composition is defined as its mass per unit volume and has units of kilograms per cubic meter (kg/m^3) in the SI system.

The **specific gravity** of a substance is a dimensionless quantity which is the ratio of the density of the substance to the density of water.

The **pressure,** P, in a fluid is the force per unit area that the fluid exerts on an object immersed in the fluid.

9.4   Variation of Pressure with Depth

In a fluid at rest, all points at the same depth are at the same pressure. **Pascal's law** states that a change in pressure applied to an enclosed fluid is transmitted undiminished to every point in the fluid and the walls of the containing vessel. *The pressure, P, at a depth of h below the surface of a liquid with density ρ and open to the atmosphere is greater than atmospheric pressure by an amount $\rho g h$.*

9.5   Pressure Measurements

The **absolute pressure** of a fluid is the sum of the **gauge pressure** and **atmospheric pressure.** The SI unit of pressure is the Pascal (Pa). Note that $1\,Pa \equiv 1\,N/m^2$.

9.6   Buoyant Forces and Archimedes's Principle

Any object partially or completely submerged in a fluid experiences a **buoyant force** equal in magnitude to the weight of the fluid displaced by the object and acting vertically upward through the point which was the center of gravity of the displaced fluid.

9.7   Fluids in Motion

Many features of fluid motion can be understood by considering the behavior of an ideal fluid, which satisfies the following conditions:

- **The fluid is nonviscous;** that is, there is no internal friction force between adjacent fluid layers.

- **The fluid is incompressible,** which means that its density is constant.

- **The fluid motion is steady,** meaning that the velocity, density, and pressure at each point in the fluid do not change in time.

- **The fluid moves without turbulence.** This implies that each element of the fluid has zero angular velocity about its center; that is, there can be no eddy currents present in the moving fluid.

Fluids which have the "ideal" properties stated above obey two important equations:

- The **equation of continuity** states that the flow rate through a pipe is constant (i.e. the product of the cross-sectional area of the pipe and the speed of the fluid is constant).

- **Bernoulli's equation** states that the sum of the pressure (P), kinetic energy per unit volume ($\rho v^2/2$), and the potential energy per unit volume ($\rho g h$) has a constant value at all points along a streamline.

9.9   Surface Tension, Capillary Action, and Viscous Fluid Flow

The concept of **surface tension** can be thought of as the energy content of the fluid at its surface per unit surface area. In general, any equilibrium configuration of an object is one in which the energy is minimum. For a given volume, the spherical shape is the one that has the smallest surface area; therefore, a drop of water takes on a spherical shape. The surface tension of liquids decreases with increasing temperature.

Forces between like molecules, such as the forces between water molecules, are called **cohesive forces** and forces between unlike molecules, such as those of glass on water are **adhesive forces.** If a capillary tube is inserted into a fluid for which adhesive forces dominate over cohesive forces, the surrounding liquid will rise into the tube. If a capillary tube is inserted into a liquid in which cohesive forces dominate over adhesive forces, the level of the liquid in the capillary tube will be below the surface of the surrounding fluid.

Viscosity refers to the internal friction of a fluid. At sufficiently high velocities, fluid flow changes from simple streamline flow to turbulent flow. The onset of turbulence in a tube is determined by a factor called the **Reynolds number,** which is a function of the density of the fluid, the average speed of the fluid along the direction of flow, the diameter of the tube, and the viscosity of the fluid.

9.10 Transport Phenomena

The two fundamental processes involved in fluid transport resulting from concentration differences are called **diffusion** and **osmosis.** In a diffusion process, molecules move from a region where their concentration is high to a region where their concentration is lower. Diffusion occurs readily in air; the process also occurs in liquids and, to a lesser extent, in solids. Osmosis is defined as the movement of water from a region where its concentration is high, across a selectively permeable membrane, into a region where its concentration is lower. Osmosis is often described simply as the diffusion of water across a membrane.

EQUATIONS AND CONCEPTS

A body of matter can be deformed (experience change in size or shape) by the application of external forces. **Stress** is a quantity which is proportional to the force which causes the deformation; strain is a measure of the degree of deformation. The **elastic modulus** is a general characterization of the deformation. *Each type of deformation is characterized by a specific modulus.*

$$\text{stress} = \text{elastic modulus} \times \text{strain} \qquad (9.1)$$

The SI **units of pressure** are newtons per square meter, or Pascals (Pa).

$$1\,\text{Pa} \equiv 1\,\text{N/m}^2$$

Young's modulus is a measure of the resistance of a body to elongation. It is defined as the ratio of tensile stress to tensile strain.

$$\frac{F}{A} = Y\frac{\Delta L}{L_0} \qquad (9.3)$$

Within a limited range of values, the graph of stress vs. strain for a given substance will be a straight line. When the stress exceeds the **elastic limit** (at the yield point), the stress-strain curve will no longer be linear.

Experimental observations

The **Shear modulus** is a measure of the deformation which occurs when a force is applied along a direction parallel to one surface of a body.

$$\frac{F}{A} = S\frac{\Delta x}{h} \tag{9.4}$$

Consider a rectangular block of height h and top and bottom surfaces each of area A. A force $\vec{F}$ applied parallel to the top surface, with the bottom surface fixed, will cause the top surface to move forward a distance Δx. *The ratio F/A is the shear stress and the ratio $\Delta x/h$ is the shear strain.*

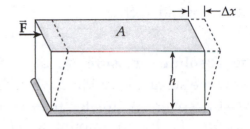

The **Bulk modulus** characterizes the response of a body to uniform pressure (or squeezing) on all sides. Note that when ΔP is positive (increase in pressure), the ratio $\Delta V/V$ will be negative (decrease in volume) and vice versa. Therefore, the negative sign in the equation ensures that B will always be positive.

$$\Delta P = -B\frac{\Delta V}{V} \tag{9.5}$$

The **density** of a homogeneous substance is defined as its ratio of mass per unit volume. *The value of density is characteristic of a particular type of material and independent of the total quantity of material in the sample.*

$$\rho \equiv \frac{M}{V} \tag{9.6}$$

The **SI units of density** are kg per cubic meter.

$$1\,\text{g}/\text{cm}^3 = 1\,000\,\text{kg}/\text{m}^3$$

The (average) **pressure** is defined as the normal force per unit area acting on a surface. Pressure has units of pascals (newtons per square meter).

$$P \equiv \frac{F}{A} \tag{9.7}$$

Atmospheric pressure is often expressed in other units:

atmosphere:

mm of mercury (Torr):

pounds per sq. inch:

Conversion of Units

$1\,\text{atm} = 1.013 \times 10^5\,\text{Pa}$

$1\,\text{Torr} = 133.3\,\text{Pa}$

$1\,\text{lb/in}^2 = 6.895 \times 10^3\,\text{Pa}$

The **absolute pressure**, P, at a depth, h, below the surface of a liquid which is open to the atmosphere is greater than atmospheric pressure, P_0, by an amount which depends on the depth below the surface. *The pressure at a depth h below the surface of a liquid does not depend on the shape of the container.*

$$P = P_0 + \rho g h \qquad (9.11)$$

$$P_0 = 1.013 \times 10^5\,\text{Pa}$$

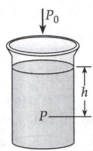

The quantity $\rho g h$ is called the **gauge pressure** and P is the absolute pressure. Therefore,

$$\begin{array}{c} \text{absolute} \\ \text{pressure} \end{array} = \begin{array}{c} \text{atmospheric} \\ \text{pressure} \end{array} + \begin{array}{c} \text{gauge} \\ \text{pressure} \end{array}$$

Fluid pressure below the surface of a liquid

Pascal's law states that pressure applied to an enclosed fluid (liquid or gas) is transmitted undiminished to every point within the fluid and over the walls of the vessel which contain the fluid.

Archimedes's principle states that when an object is partially or fully submerged in a fluid, the fluid exerts an upward **buoyant force**, B, which depends on the fluid density and the displaced volume. *The buoyant force equals the weight of the displaced fluid.*

$$B = \rho_{\text{fluid}} V_{\text{fluid}} g \qquad (9.12b)$$

V_{fluid} = volume of displaced fluid

Fluid dynamics, the treatment of fluids in motion, is greatly simplified under the assumption that the fluid is ideal with the following characteristics:

- **nonviscous** — internal friction between adjacent fluid layers is negligible

- **incompressible** — the density throughout the fluid is constant.

- **steady** — the velocity, density, and pressure at each point in the fluid are constant in time

- **irrotational** (without turbulence) — there are no eddy currents within the fluid (each element of the fluid has zero angular velocity about its center)

Properties of ideal fluids

For an **incompressible fluid** (ρ = constant), the equation of continuity can be written as Equation 9.15. The product Av is called the flow rate. *The **flow rate** at any point along a pipe carrying an incompressible fluid is constant.*

$$A_1 v_1 = A_2 v_2 \tag{9.15}$$

Continuity equation for incompressible fluid

Bernoulli's equation is the most fundamental law in fluid mechanics. It is a statement of the law of conservation of mechanical energy as applied to a fluid. Bernoulli's equation states that the sum of pressure, kinetic energy per unit volume, and potential energy per unit volume remains constant along a streamline of an ideal fluid.

$$P + \tfrac{1}{2}\rho v^2 + \rho g y = \text{constant} \tag{9.17}$$

The **surface tension**, γ, in a film of liquid is defined as the ratio of the magnitude of the surface tension force, F, to the length, L, along which the force acts. Note: If F is the surface tension force required to lift a small wire from the surface of a liquid, the value L in Equation 9.19 must be twice the length of the wire because the force of surface tension acts along both sides of the wire as it is being lifted from the liquid.

$$\gamma \equiv \frac{F}{L} \qquad (9.19)$$

In a **capillary tube**, the angle between the solid surface and a line drawn tangent to the liquid at the surface is called the **angle of contact**. Note that ϕ is less than 90° for any substance in which adhesive forces are stronger than cohesive forces. In such a case, the liquid will rise to a height h in the tube. If cohesive forces dominate, the liquid in the tube will be depressed below the surface of the surrounding liquid by a distance h.

$$h = \frac{2\gamma}{\rho g r} \cos\phi \qquad (9.22)$$

Consider a layer of liquid of thickness d and area A between two solid surfaces. The **coefficient of viscosity**, η (the lowercase Greek letter, eta), for the fluid is defined in terms of the force, F, required to move one of the solid surfaces with a velocity v. The coefficient of viscosity may also be thought of as the ratio of the shearing stress to the rate of change of the shear strain. The SI units of viscosity are N·s/m².

$$F\eta = \frac{Av}{d} \qquad (9.23)$$

$$1 \text{ poise} = 10^{-1} \text{ N·s/m}^2 \qquad (9.24)$$

The onset of **turbulence** in a tube is determined by a dimensionless factor called the **Reynolds number**, which is a function of the density of the fluid, the average speed of the fluid along the direction of flow, the diameter of the tube, and the viscosity of the fluid. In the region between 2000 and 3000, the flow is unstable, meaning that any small disturbance will cause its motion to change from streamline to turbulent flow.

$$RN = \frac{\rho v d}{\eta} \qquad (9.26)$$

Streamline Flow: $RN < {\sim}2\,000$

Unstable Flow: $2\,000 < RN < 3\,000$

Turbulent Flow: $RN > 3\,000$

The basic equation for diffusion is Fick's law. The **diffusion rate**, a measure of the mass being transported per unit time, is proportional to the cross-sectional area A and to the change in concentration per unit distance, $(C_2 - C_1)/L$, which is called the concentration gradient. The proportionality constant D is called the **diffusion coefficient** and has units of square meters per second.

$$\text{Diffusion Rate} = \frac{\Delta M}{\Delta t} = DA\left(\frac{C_2 - C_1}{L}\right) \qquad (9.27)$$

D = diffusion coefficient

Stokes' law describes the resistive force on a small spherical object of radius r falling with speed v through a viscous fluid.

$$F_r = 6\pi \eta r v \qquad (9.28)$$

As a sphere falls through a viscous medium, three forces act on it: the force of frictional resistance, the buoyant force of the fluid, and the weight of the sphere. When the net upward force balances the downward weight force, the sphere reaches **terminal speed**.

$$v_t = \frac{2r^2 g}{9\eta}\left(\rho - \rho_f\right) \qquad (9.29)$$

In a **centrifuge** those particles having the greatest mass will have the largest terminal speed. Therefore, the most massive particles will settle out of the mixture. The factor of k is a coefficient of frictional resistance which must be determined experimentally.

$$v_t = \frac{m\omega^2 r}{k}\left(1 - \frac{\rho_f}{\rho}\right) \qquad (9.32)$$

REVIEW CHECKLIST

▷ Describe the three types of deformations that can occur in a solid, and define the elastic modulus that is used to characterize each: (Young's modulus, Shear modulus, and Bulk modulus).

▷ Understand the concept of pressure at a point in a fluid, and the variation of pressure with depth. Understand the relationships among absolute, gauge, and atmospheric pressure values; and know the several different units commonly used to express pressure.

▷ Understand the origin of buoyant forces; and state and explain Archimedes's principle.

▷ State and understand the physical significance of the equation of continuity (constant flow rate) and Bernoulli's equation for fluid flow (relating flow velocity, pressure, and pipe elevation).

SOLUTIONS TO SELECTED END-OF-CHAPTER PROBLEMS

4. When water freezes, it expands about 9.00%. What would be the pressure increase inside your automobile engine block if the water in it froze? The bulk modulus of ice is 2.00×10^9 N/m^2.

Solution

We assume that the coolant space inside the engine block was completely filled with water. In the liquid state, this water occupied some volume, V_{liquid}. When frozen, the water would normally occupy a volume $1.0900\,V_{liquid}$. However, unless the engine block bursts , sufficient pressure builds up to compress the ice and force it to occupy volume V_{liquid}.

Consider the increase in pressure required to compress ice into the space it would occupy in the liquid state. Then, the original volume of the ice is $V_0 = 1.0900\,V_{liquid}$, the final volume is $V_f = V_{liquid}$, and the change in volume is

$$\Delta V = V_f - V_0 = V_{liquid} - 1.0900\,V_{liquid} = -0.0900\,V_{liquid}$$

From the definition of bulk modulus, $\qquad B = -\dfrac{\Delta P}{\Delta V / V_0}$

The increase in pressure required to accomplish this compression is

$$\Delta P = -B\left(\frac{\Delta V}{V_0}\right) = -\left(2.00 \times 10^9 \text{ Pa}\right)\left(\frac{-0.0900\,V_{liquid}}{1.0900\,V_{liquid}}\right) = 1.65 \times 10^8 \text{ Pa} = 165 \text{ MPa} \qquad \lozenge$$

Note that this increase in pressure is

$$\Delta P = \left(1.65 \times 10^8 \text{ Pa}\right)\left(\frac{1 \text{ atm}}{1.013 \times 10^5 \text{ Pa}}\right) \approx 1600 \text{ atm}$$

Thus, there is a good possibility that the engine block will burst before the compression is accomplished.

17.   If 1.0 m³ of concrete weighs 5.0 × 10⁴ N, what is the height of the tallest cylindrical concrete pillar that will not collapse under its own weight? The compression strength of concrete (the maximum pressure that can be exerted on the base of the structure) is 1.7×10^7 Pa.

Solution

Note that the product $g\rho_{concrete}$, where $\rho_{concrete}$ is the mass per unit volume (density) for concrete, gives the weight per unit volume. We assume the cylindrical pillar has height h and cross-sectional area A. Hence, the volume is $V = Ah$, and the weight of the pillar will be

$$F_g = (g\rho_{concrete})V = (g\rho_{concrete})Ah$$

The pressure on the concrete forming the base of the structure is then

$$P = \frac{F_g}{A} = \frac{(g\rho_{concrete})Ah}{A} = (g\rho_{concrete})h$$

If the maximum pressure concrete can withstand is $P_{max} = 1.7 \times 10^7$ Pa, the maximum height the pillar can have without collapsing under its own weight is

$$h_{max} = \frac{P_{max}}{(g\rho_{concrete})} = \frac{1.7 \times 10^7 \text{ Pa}}{5.0 \times 10^4 \text{ N/m}^3} = 340 \text{ m} \approx 1100 \text{ ft} \qquad \lozenge$$

_____

23.   A container is filled to a depth of 20.0 cm with water. On top of the water floats a 30.0-cm-thick layer of oil with specific gravity 0.700. What is the absolute pressure at the bottom of the container?

Solution

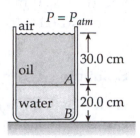

If P_0 is the known absolute pressure at some chosen reference level, the absolute pressure at a depth h, in a fluid of density ρ, below that reference level is $P = P_0 + \rho g h$ where g is the acceleration due to gravity. In this problem, there are multiple fluids involved and it may appear ambiguous as to what density (that is, density of air, oil or water) should be used in this equation for the pressure.

The answer to this dilemma is to deal with one fluid at a time. To find the absolute pressure at the bottom of the water layer (i.e., at the bottom of the container), first compute the absolute pressure at the boundary between the oil and water.

Atmospheric pressure exists at the upper surface of the oil (at the air-oil boundary). Choosing this boundary as the reference level, the absolute pressure at point A on the oil-water boundary is:

$$P_A = P_{atm} + \rho_{oil}\, g\, h_{oil}$$

The oil's specific gravity is

$$\left(s.\,g.\right)_{oil} = \frac{\rho_{oil}}{\rho_{water}} = 0.700$$

$$\rho_{oil} = \left(0.700\right)\rho_{water} = \left(0.700\right)\left(1\,000\ \text{kg/m}^3\right) = 700\ \text{kg/m}^3$$

and, $P_A = 1.013 \times 10^5\ \text{Pa} + \left(700\ \text{kg/m}^3\right)\left(9.80\ \text{m/s}^2\right)\left(0.300\ \text{m}\right) = 1.03 \times 10^5\ \text{Pa}$

Now that the pressure at point A is known, the oil-water boundary can be chosen as the new reference level (so $P_0 = P_A = 1.03 \times 10^5\ \text{Pa}$) and the absolute pressure at point B on the bottom of the container can be computed as $P_B = P_A + \rho_{water}\, g\, h_{water}$.

This yields

$$P_B = 1.03 \times 10^5\ \text{Pa} + \left(1\,000\ \text{kg/m}^3\right)\left(9.80\ \text{m/s}^2\right)\left(0.200\ \text{m}\right) = 1.05 \times 10^5\ \text{Pa} \qquad \lozenge$$

31.   A bathysphere used for deep sea exploration has a radius of 1.50 m and a mass of 1.20 × 10⁴ kg. In order to dive, the sphere takes on mass in the form of sea water. Determine the mass the bathysphere must take on so that it can descend at a constant speed of 1.20 m/s when the resistive force on it is 1 100 N upward. The density of sea water is $1.03 \times 10^3 \text{ kg/m}^3$.

Solution

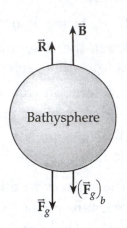

Bathysphere

As the bathysphere descends, there are four forces acting on it as shown in the diagram at the right. These forces are: (1) the upward buoyant force $\vec{B}$ exerted on the bathysphere by the water; (2) an upward resistance force $\vec{R}$ opposing the motion of the bathysphere through the water; the weight $\vec{F}_g$ of the bathysphere itself; and the weight $\left(\vec{F}_g\right)_b$ of the sea water taken on board as ballast.

When the submarine descends at constant speed, it is in equilibrium and $\Sigma F_y = 0$ gives

$$B + R - F_g - \left(F_g\right)_b = 0 \qquad \text{or} \qquad \left(F_g\right)_b = B + R - F_g$$

The buoyant force is $B = \left(\rho_{\substack{sea \\ water}} V_{\substack{water \\ displaced}} \right) g = \left(\rho_{\substack{sea \\ water}} V_{\substack{bathysphere}} \right) g = \rho_{\substack{sea \\ water}} \left(\dfrac{4\pi r^3}{3} \right) g$

Thus, $\left(F_g\right)_b = m_{ballast} g = \rho_{\substack{sea \\ water}} \left(\dfrac{4\pi r^3}{3} \right) g + R - \left(m_{bathyspheere} \right) g$

and the mass of sea water that should be taken on as ballast is

$$m_{ballast} = \rho_{\substack{sea \\ water}} \left(\frac{4\pi r^3}{3} \right) + \frac{R}{g} - m_{bathyspheere}$$

or $m_{ballast} = \left(1.03 \times 10^3 \text{ kg/m}^2\right) \dfrac{4\pi (1.50 \text{ m})^3}{3} + \left(\dfrac{1\,100 \text{ N}}{9.80 \text{ m/s}^2} \right) - 1.20 \times 10^4 \text{ kg}$

yielding $m_{ballast} = 2.67 \times 10^3 \text{ kg}$ ◊

39. A 1.00-kg beaker containing 2.00 kg of oil (density $= 916$ kg/m^3) rests on a scale. A 2.00-kg block of iron is suspended from a spring scale and is completely submerged in the oil (Fig. P9.39). Find the equilibrium readings of both scales.

Solution

The volume of the iron block is

$$V = \frac{m_{iron}}{\rho_{iron}} = \frac{2.00 \text{ kg}}{7.86 \times 10^3 \text{ kg/m}^3} = 2.54 \times 10^{-4} \text{ m}^3$$

Figure P9.39

and the buoyant force exerted on the iron by the oil is

$$B = (\rho_{oil}V)g = (916 \text{ kg/m}^3)(2.54 \times 10^{-4} \text{ m}^3)(9.80 \text{ m/s}^2) = 2.28 \text{ N}$$

Applying $\Sigma F_y = 0$ to the iron block gives the support force exerted by the upper spring scale (and hence the reading on that scale) as

$$F_{upper} = m_{iron}g - B = 19.6 \text{ N} - 2.28 \text{ N} = 17.3 \text{ N}$$

From Newton's third law, the iron exerts a reaction force of magnitude B downward on the oil (and hence the beaker). Other vertical forces acting on the beaker are (1) the combined weight of the beaker and oil, and (2) an upward support force exerted by the lower scale. Applying $\Sigma F_y = 0$ to the system consisting of the beaker and the oil gives

$$F_{lower} - B - (m_{oil} + m_{beaker})g = 0$$

The support force exerted by the lower scale (and the lower scale reading) is then

$$F_{lower} = B + (m_{oil} + m_{beaker})g = 2.28 \text{ N} + \left[(2.00 + 1.00) \text{ kg}\right](9.80 \text{ m/s}^2) = 31.7 \text{ N}$$

45.   A jet of water squirts out horizontally from a hole near the bottom of the tank shown in Figure P9.45. If the hole has a diameter of 3.50 mm, what is the height, h, of the water level in the tank?

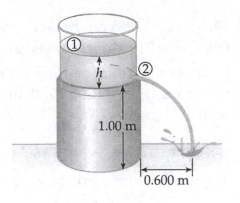

Figure P9.45

Solution

First, consider the projectile motion of the water droplets as they go from point 2 to the ground. From $\Delta y = v_{0y}t + \frac{1}{2}a_y t^2$ with $v_{0y} = 0$, we find the time of flight as

$$t = \sqrt{\frac{2(\Delta y)}{a_y}} = \sqrt{\frac{2(-1.00 \text{ m})}{-9.80 \text{ m/s}^2}} = 0.452 \text{ s}$$ ◊

The speed of the water as it emerges from the hole at point 2 may be determined from the horizontal motion:

$$v_2 = v_{0x} = \frac{\Delta x}{t} = \frac{0.600 \text{ m}}{0.452 \text{ s}} = 1.33 \text{ m/s}$$

We now use Bernoulli's equation, $P_1 + \frac{1}{2}\rho v_1^2 + \rho g y_1 = P_2 + \frac{1}{2}\rho v_2^2 + \rho g y_2$, with point 1 at the upper surface of the water in the tank and point 2 at the hole. The tank is open to the atmosphere at both points 1 and 2, so $P_1 = P_2 = P_{atm}$.

We assume the tank is large enough that the speed of the water is negligible at point 1 in comparison to the speed of the emerging water (that is, $v_1 \cong 0$). Then, Bernoulli's equation reduces to

$$\rho g y_1 = \frac{1}{2}\rho v_2^2 + \rho g y_2$$

and the height, h, of the water level in the tank is

$$h = y_1 - y_2 = \frac{v_2^2}{2g} = \frac{(1.33 \text{ m/s})^2}{2(9.80 \text{ m/s}^2)} = 9.00 \times 10^{-2} \text{ m} = 9.00 \text{ cm}$$ ◊

Note that neither the diameter of the hole nor the fact that the fluid in the tank is water is needed for this solution.

49. Old Faithful Geyser in Yellowstone Park erupts at approximately 1-hour intervals, and the height of the fountain reaches 40.0 m. (a) Consider the rising stream as a series of separate drops. Analyze the free-fall motion of one of the drops to determine the speed at which the water leaves the ground. (b) Treat the rising stream as an ideal fluid in streamline flow. Use Bernoulli's equation to determine the speed of the water as it leaves ground level. (c) What is the pressure (above atmospheric pressure) in the heated underground chamber 175 m below the vent? You may assume that the chamber is large compared with the geyser vent.

Solution

(a) Consider the motion of a water-drop projectile as it goes from the geyser vent to the top of the fountain. We use $v_y^2 = v_{0y}^2 + 2a_y \Delta y$, with $v_y = 0$ when Δy equals the height h of the fountain. Then, the speed of the drop as it emerges from the vent is

$$v_{vent} = v_{0y} = \sqrt{v_y^2 - 2(-g)h} = \sqrt{0 - 2(-9.80 \text{ m/s}^2)(40.0 \text{ m})} = 28.0 \text{ m/s} \qquad \lozenge$$

(b) Because of the low density of air and the small change in altitude, we neglect any change in atmospheric pressure in going from the ground to the top of the fountain. Then, applying Bernoulli's equation from the vent to the top of the fountain gives

$$P_{vent} + \tfrac{1}{2}\rho_{water}v_{vent}^2 + \rho_{water}gy_{vent} = P_{top} + \tfrac{1}{2}\rho_{water}v_{top}^2 + \rho_{water}gy_{top}$$

Thus,
$$v_{vent} = \sqrt{v_{top}^2 + \frac{2}{\rho_{water}}\left[\left(P_{top} - P_{vent}\right) + \rho_{water}g\left(y_{top} - y_{vent}\right)\right]}$$

with $P_{top} \approx P_{vent}$, $v_{top} = 0$, and $\left(y_{top} - y_{vent}\right) = h$, or

$$v_{vent} = \sqrt{2gh} = \sqrt{2(9.80 \text{ m/s}^2)(40.0 \text{ m})} = 28.0 \text{ m/s} \qquad \lozenge$$

(c) If the chamber may be considered large in comparison to the geyser vent, we may assume that the speed of the water in the chamber is negligible in comparison to its speed at the geyser vent. Then applying Bernoulli's equation between the chamber and geyser vent gives

$$P_{chamber} + 0 + \rho_{water}gy_{chamber} = P_{vent} + \tfrac{1}{2}\rho_{water}v_{vent}^2 + \rho_{water}gy_{vent}$$

With $P_{vent} = P_{atmo}$, this gives the gauge pressure in the chamber as

$$P_{gauge} = \left(P_{chamber} - P_{atmo}\right) = \rho_{water}\left[\tfrac{1}{2}v_{vent}^2 + g\left(y_{vent} - y_{chamber}\right)\right]$$

or
$$P_{gauge} = (1.00\times10^3 \text{ kg/m}^3)\left[\tfrac{1}{2}(28.0 \text{ m/s})^2 + (9.80 \text{ m/s}^2)(175 \text{ m})\right]$$

This yields $P_{gauge} = 2.11\times10^6$ Pa = 2.11 MPa = 20.8 atmospheres $\qquad \lozenge$

54.   Whole blood has a surface tension of 0.058 N/m and a density of $1\,050 \text{ kg/m}^3$. To what height can whole blood rise in a capillary blood vessel that has a radius of 2.0×10^{-6} m if the contact angle is zero?

Solution

The surface tension γ of a fluid is defined as the tension force per unit length (tangential to the fluid surface and tending to cause that surface to contract) along any line drawn on the surface of the fluid.

Consider the line along which the upper surface of a fluid in a capillary tube meets the wall of that tube as shown in the sketch. This line has a length equal to the circumference of the tube (that is, $L=2\pi r$). The surface (and hence the tension force) makes an angle ϕ, known as the contact angle, with the vertical wall of the tube at points on this line. The total **upward** (vertical) force the tube wall exerts on the fluid is then

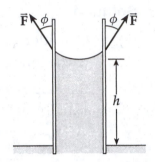

$$F_v = F\cos\phi = (\gamma L)\cos\phi = 2\pi\gamma\, r\cos\phi$$

The fluid then rises until the weight of the fluid lifted, $w = mg = \rho V g = \rho\left(\pi r^2 h\right)g$, equals the upward force F_v. When equilibrium is reached, we then have

$$\rho\left(\pi r^2 h\right)g = 2\pi\gamma\, r\cos\phi \qquad \text{or} \qquad h = \frac{2\gamma\cos\phi}{\rho g r} \qquad \text{[Equation 9.22]}$$

If the contact angle is zero when whole blood rises in a capillary blood vessel, the height to which blood will rise in a vessel with a radius of $r = 2.0\times10^{-6}$ m is:

$$h = \frac{2(0.058 \text{ N/m})\cos 0°}{\left(1\,050 \text{ kg/m}^3\right)\left(9.80 \text{ m/s}^2\right)\left(2.0\times10^{-6} \text{ m}\right)} = 5.6 \text{ m} \qquad\qquad \Diamond$$

61.   Spherical particles of a protein of density 1.8 g/cm^3 are shaken up in a solution of 20°C water. The solution is allowed to stand for 1.0 h. If the depth of water in the tube is 5.0 cm, find the radius of the largest particles that remain in solution at the end of the hour.

Solution

Note that the density of the protein is

$$\rho = 1.8 \ \frac{g}{cm^3}\left(\frac{1 \ kg}{10^3 \ g}\right)\left(\frac{10^2 \ cm}{1 \ m}\right)^3 = 1.8 \times 10^3 \ kg/m^3$$

If a particle is still in suspension after one hour, then its terminal speed must be

$$v_t \le \left(\frac{5.0 \ cm}{1.0 \ h}\right)\left(\frac{1 \ m}{100 \ cm}\right)\left(\frac{1 \ h}{3600 \ s}\right) = 1.4 \times 10^{-5} \ m/s$$

In general, the terminal speed of a particle of density ρ and radius r falling through a fluid of density ρ_f and viscosity η is given by

$$v_t = \frac{2r^2 g}{9\eta}(\rho - \rho_f)$$

Thus, if the upper limit of the terminal velocity of the particles still suspended in the water is $v_{t, \ max} = 1.4 \times 10^{-5} \ m/s$, the maximum radius of these particles is

$$r_{max} = \sqrt{\frac{9\eta \ v_{t, \ max}}{2g(\rho - \rho_f)}} = \sqrt{\frac{9(1.0 \times 10^{-3} \ N \cdot s/m^2)(1.4 \times 10^{-5} \ m/s)}{2(9.8 \ m/s^2)[(1.8 - 1.0) \times 10^3 \ kg/m^3]}}$$

or $r_{max} = 2.8 \times 10^{-6} \ m = 2.8 \ \mu m$ ◊

73.   The approximate inside diameter of the aorta is 0.50 cm; that of a capillary is 10 μm. The approximate average blood flow speed is 1.0 m/s in the aorta and 1.0 cm/s in the capillaries. If all the blood in the aorta eventually flows through the capillaries, estimate the number of capillaries in the circulatory system.

Solution

If the diameters of the aorta and of a capillary are d_1 and d_2, respectively,

their cross-sectional areas are $$A_{aorta} = A_1 = \frac{\pi d_1^2}{4}$$

and $$A_{capillary} = A_c = \frac{\pi d_2^2}{4}$$

Assuming the circulatory system has a total of N capillaries, the total cross-sectional area carrying blood from the aorta is $A_2 = NA_{capillary} = N\left(\pi d_2^2\right)/4$.

The equation of continuity then requires that $A_2 v_2 = A_1 v_1$, where v_1 is the blood flow speed in the aorta and v_2 is the flow speed in a capillary.

This gives $$N\left(\frac{\pi d_2^2}{4}\right)v_2 = \left(\frac{\pi d_1^2}{4}\right)v_1$$

so the number of capillaries in the circulatory system must be

$$N = \left(\frac{d_1}{d_2}\right)^2\left(\frac{v_1}{v_2}\right) = \left(\frac{0.50\times10^{-2} \text{ m}}{10\times10^{-6} \text{ m}}\right)^2\left(\frac{1.0 \text{ m/s}}{1.0\times10^{-2} \text{ m/s}}\right) = 2.5\times10^7 = 25 \text{ million} \qquad \Diamond$$

79. A block of wood weighs 50.0 N in air. A sinker is attached to the block, and the weight of the wood-sinker combination is 200 N when the sinker alone is immersed in water. Finally, the wood-sinker combination is completely immersed, and the weight is measured to be 140 N. Find the density of the block.

Solution

In the sketches below, $T_1 = 200$ N is the scale reading when only the sinker is submerged and $T_2 = 140$ N is the scale reading with both the sinker and block submerged. $\vec{B}_s$ is the buoyant force exerted on the sinker by the water, while $\vec{B}_b$ is the buoyant force the water exerts on the block.

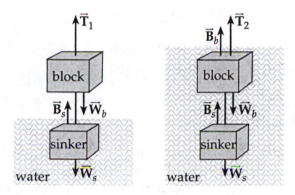

Applying the first condition of equilibrium to the entire system shown in each case gives

With only sinker submerged: $\qquad \Sigma F_y = 0 \Rightarrow T_1 = (W_b + W_s) - B_s$ [1]

With both block and sinker submerged: $\quad \Sigma F_y = 0 \Rightarrow T_2 = (W_b + W_s) - B_s - B_b$ [2]

Subtracting the second of these equations from the first gives: $\quad B_b = T_1 - T_2$

From Archimedes's principle, $B_b = \rho_{water} g V_b$. Thus, the volume of the wooden block is $V_b = (T_1 - T_2)/(\rho_{water} g)$, while its mass is $m_b = W_b/g = (50.0 \text{ N})/g$. The density of the wood is therefore

$$\rho_{wood} = \frac{m_b}{V_b} = \frac{(50.0 \text{ N})\rho_{water}}{(T_1 - T_2)} = \frac{(50.0 \text{ N})(1.00 \times 10^3 \text{ kg/m}^3)}{200 \text{ N} - 140 \text{ N}} = 833 \text{ kg/m}^3 \qquad \Diamond$$

85.   A 2.0-cm-thick bar of soap is floating on a water surface so that 1.5 cm of the bar is under water. Bath oil of specific gravity 0.60 is poured into the water and floats on top of it. What is the depth of the oil layer when the top of the soap is just level with the upper surface of the oil?

Solution

When the soap bar floats in water alone, it floats to a depth of 1.5 cm and displaces a volume of water equal to $V_1 = A(1.5 \text{ cm})$ where A is the surface area of either the top or bottom of the rectangular bar. Since the bar is floating, its weight must equal the buoyant force exerted by the water , or

$$w_{bar} = B_1 = \rho_{water} g V_1 = \rho_{water} g A(1.5 \text{ cm})$$

When the bar floats in the oil-water combination as shown at the right, the total buoyant force must equal the weight of the bar , or

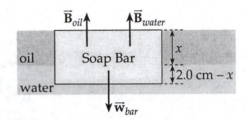

$$B_{oil} + B_{water} = w_{bar} = \rho_{water} g A(1.5 \text{ cm})$$

From Archimedes's principle, $B_{oil} = \rho_{oil} g A x$ and $B_{water} = \rho_{water} g A(2.0 \text{ cm} - x)$

We now have

$$\rho_{oil} g A x + \rho_{water} g A(2.0 \text{ cm} - x) = \rho_{water} g A(1.5 \text{ cm})$$

or, solving for x:

$$x = \left(\frac{\rho_{water}}{\rho_{water} - \rho_{oil}} \right)(2.0 \text{ cm} - 1.5 \text{ cm}) = \frac{2.0 \text{ cm} - 1.5 \text{ cm}}{1 - \rho_{oil}/\rho_{water}}$$

But, the specific gravity of the oil is s.g. of oil $= \rho_{oil}/\rho_{water} = 0.60$, so the thickness of the oil layer is

$$x = \frac{2.0 \text{ cm} - 1.5 \text{ cm}}{1 - 0.60} = 1.3 \text{ cm}$$ ◊

Chapter 10
Thermal Physics

NOTES FROM SELECTED CHAPTER SECTIONS

10.1 Temperature and the Zeroth Law of Thermodynamics

The zeroth law of thermodynamics (or the equilibrium law) can be stated as follows:

If bodies A and B are separately in thermal equilibrium with a third body, C, then A and B will be in thermal equilibrium with each other if placed in thermal contact.

Two objects in thermal equilibrium with each other are at the same temperature.

10.2 Thermometers and Temperature Scales

The physical property used in a constant volume gas thermometer is the pressure variation with temperature of a fixed volume of gas. The temperature readings are nearly independent of the substance used in the thermometer.

The **triple point of water**, which is the single temperature and pressure at which water, water vapor, and ice can coexist in equilibrium, was chosen as a convenient and reproducible reference temperature for the Kelvin scale. It occurs at a temperature of 0.01 °C and a pressure of 4.58 mm of mercury. The temperature at the triple point of water on the Kelvin scale has been assigned a value of 273.16 kelvins (K). Thus, the SI unit of temperature, the kelvin, is defined as 1/273.16 of the temperature of the triple point of water.

10.3 Thermal Expansion of Solids and Liquids

Liquids generally increase in volume with increasing temperature and have volume expansion coefficients about ten times greater than those of solids. Water is an exception to this rule; as the temperature increases from 0 °C to 4 °C, water contracts and thus its density increases. Above 4 °C, water expands with increasing temperature. *The density of water reaches its maximum value ($1\,000\ kg/m^3$) 4 degrees above the freezing point.*

10.4   Macroscopic Description of an Ideal Gas

In an **ideal gas** the atoms or molecules move randomly, the individual particles do not exert long-range forces on each other, and each particle is considered to be a "point" mass. One **mole** of any substance is that quantity of material that contains a number of particles equal to the number of atoms in 12 grams of the isotope carbon 12. **One-mole quantities of all gases at standard temperature and pressure contain the same numbers of molecules.** *Equal volumes of gas at the same temperature and pressure contain the same numbers of molecules. There are* 6.02×10^{23} *(Avogadro's number, N_A) particles in one mole of any element.*

10.5   The Kinetic Theory of Gases

A microscopic **model of an ideal gas** is based on the following assumptions:

- **The number of molecules is large, and the average separation between them is large** compared with their dimensions. Therefore, the molecules occupy a negligible volume compared with the volume of the container.

- **The molecules obey Newton's laws of motion, but the individual molecules move in a random fashion.** By random fashion, we mean that the molecules move in all directions with equal probability and with various speeds. This distribution of velocities does not change in time, despite the collisions between molecules.

- **The molecules undergo elastic collisions with each other.** Thus, the molecules are considered to be structureless (that is, point masses), and in the collisions both kinetic energy and momentum are conserved.

- **The forces between molecules are negligible, except during a collision.** The forces between molecules are short-range, so that the only time the molecules interact with each other is during a collision.

- **The gas under consideration is a pure gas.** That is, all molecules are identical.

- **The molecules of the gas make perfectly elastic collisions with the walls of the container.** Hence, the wall will eject as many molecules as it absorbs, and the ejected molecules will have the same average kinetic energy as the absorbed molecules.

EQUATIONS AND CONCEPTS

T_C is the Celsius temperature and T is the Kelvin temperature (sometimes called the absolute temperature). The size of a degree on the Kelvin scale is identical to the size of a degree on the Celsius scale.

$$T_C = T - 273.15 \tag{10.1}$$

Equations 10.2a and 10.2b are useful in **converting temperature values** between Fahrenheit and Celsius scales.

$$T_F = \tfrac{9}{5}T_C + 32 \tag{10.2a}$$

$$T_C = \tfrac{5}{9}(T_F - 32) \tag{10.2b}$$

The SI unit of temperature is the **kelvin**, K, which is defined as 1/273.16 of the temperature of the triple point of water. The notations °C and °F refer to actual temperature values in degrees Celsius and degrees Fahrenheit.

Units and notation

These are two forms of the basic equation for the **thermal expansion** of a solid. The change in length is proportional to the original length and to the change in temperature. The constant α is characteristic of a particular type of material and is called the average temperature coefficient of linear expansion.

$$\Delta L = \alpha L_0 \Delta T \tag{10.4}$$

$$L = L_0(1 + \alpha \Delta T)$$

α = coefficient of linear expansion

If a body's temperature changes, its surface area and its volume will change by amounts which are proportional to the changes in temperature. γ (gamma) is the average temperature coefficient of area expansion; β (beta) is the average temperature coefficient of volume expansion.

$$\Delta A = \gamma A_0 \Delta T \tag{10.5}$$

$$\Delta V = \beta V_0 \Delta T \tag{10.6}$$

$\gamma \cong 2\alpha$ = coefficient of area expansion

$\beta \cong 3\alpha$ = coefficient of volume expansion

The **number of moles** in a sample of an element or compound is the ratio of the mass, m, of the sample to the atomic or molar mass, of the material. *Also, one mole of a substance contains Avogadro's number of molecules.*

$$n = \frac{m}{\text{molar mass}} \quad (10.7)$$

$$N_A = 6.02 \times 10^{23} \, \frac{\text{particles}}{\text{mole}}$$

$$m_{\text{atom}} = \frac{\text{molar mass}}{N_A}$$

Boyle's law states that when a gas is maintained at constant temperature, its pressure is inversely proportional to its volume.

$$P \propto \left(\frac{1}{V} \right) \text{ (when } T = \text{constant)}$$

Charles' law states that when constant pressure is maintained, a gas's volume is directly proportional to its absolute temperature.

$$V \propto T \text{ (when } P = \text{constant)}$$

Gay-Lussac's law states that when the volume is held constant, the pressure is directly proportional to the absolute temperature.

$$P \propto T \text{ (when } V = \text{constant)}$$

This is the **equation of state of an ideal gas**. In this equation, T must be the temperature in kelvins. In this equation n is the number of moles of gas in the sample and R is the universal gas constant. R must be expressed in units which are consistent with those used for pressure P and volume V.

$$PV = nRT \quad (10.8)$$

n= number of moles in sample

$$R = 8.31 \, \text{J/mol} \cdot \text{K} \quad (10.9)$$

$$R = 0.0821 \, \text{L} \cdot \text{atm/mol} \cdot \text{K}$$

The **number of moles** in a sample of gas equals the number of molecules, N, divided by Avogadro's number, N_A.

$$n = \frac{N}{N_A} \quad (10.10)$$

The **ideal gas law** can also be expressed in this alternate form where N is the total number of molecules in the sample and k_B is Boltzmann's constant.

$$PV = Nk_B T \quad (10.11)$$

N=number of molecules in sample

$$k_B = \frac{R}{N_A} = 1.38 \times 10^{-23} \, \text{J/K} \quad (10.12)$$

The **pressure of an ideal gas** is proportional to the number of molecules per unit volume and proportional to the average kinetic energy of the molecules.

$$P = \frac{2}{3}\left(\frac{N}{V}\right)\left(\frac{1}{2}m\overline{v^2}\right) \qquad (10.13)$$

The **average translational kinetic energy** per molecule is directly proportional to the absolute temperature of the gas.

$$\frac{1}{2}m\overline{v^2} = \frac{3}{2}k_B T \qquad (10.15)$$

This expression for the **root mean square (rms) speed** shows that, at a given temperature, lighter molecules move faster on the average than heavier ones.

$$v_{rms} = \sqrt{\overline{v^2}} = \sqrt{\frac{3k_B T}{m}} = \sqrt{\frac{3RT}{M}} \qquad (10.18)$$

m = molecular mass

M = molar mass

Review Checklist

▷ Describe the operation of the constant-volume gas thermometer and how it is used to determine the Kelvin temperature scale. Convert between the various temperature scales, especially the conversion from degrees Celsius into kelvins, degrees Fahrenheit into kelvins, and degrees Celsius into degrees Fahrenheit.

▷ Define the linear expansion coefficient and volume expansion coefficient for an isotropic solid, and understand how to use these coefficients in practical situations involving expansion or contraction.

▷ Understand the assumptions made in developing the molecular model of an ideal gas; and apply the equation of state for an ideal gas to calculate pressure, volume, temperature, or number of moles.

▷ Define each of the following terms: molecular weight, mole, Avogadro's number, universal gas constant, and Boltzman's constant.

SOLUTIONS TO SELECTED END-OF-CHAPTER PROBLEMS

1.   For each of the following temperatures, find the equivalent temperature on the indicated scale: (a) –273.15°C on the Fahrenheit scale, (b) 98.6°F on the Celsius scale, and (c) 100 K on the Fahrenheit scale.

Solution

(a)   Note that the given temperature, $T = -273.15°C = 0$ K, corresponds to absolute zero, the temperature at which an ideal gas will exert zero pressure. To convert this to the Fahrenheit scale, we use $T_F = \frac{9}{5}T_C + 32$ and find that absolute zero on the Fahrenheit scale occurs at

$$T_F = \frac{9}{5}(-273.15) + 32.00 = -459.7°F \approx -460°F \qquad \Diamond$$

(b)   In this case, the given temperature, $T_F = 98.6°F$, is normal body temperature. On the Celsius scale, this temperature is

$$T_C = \frac{5}{9}(T_F - 32) = \frac{5}{9}(98.6 - 32.0) = 37.0°C \qquad \Diamond$$

(c)   To convert a temperature on the Kelvin scale to the Fahrenheit scale, we first convert to the Celsius scale. Then, we convert the temperature from the Celsius to the Fahrenheit scale.

If $T = 100$ K, the corresponding temperature on the Celsius scale is

$$T_C = T - 273.15 = 100 - 273.15 = -173.15°C$$

Then, the corresponding temperature on the Fahrenheit scale is found to be

$$T_F = \frac{9}{5}T_C + 32 = \frac{9}{5}(-173.15) + 32 = -280°F \qquad \Diamond$$

5.   Show that the temperature –40° is unique in that it has the same numerical value on the Celsius and Fahrenheit scales.

Solution

The conversion from Celsius temperatures to the corresponding Fahrenheit temperature is $T_F = \dfrac{9}{5} T_C + 32$.

Observe that for a fairly high Celsius temperature, such as $T_C = 100°C$, the corresponding Fahrenheit temperature is $T_F = \dfrac{9}{5}(100) + 32 = 212°F$ or the numeric value of the Fahrenheit scale reading is greater than the corresponding reading on the Celsius scale. On the other hand, if the Celsius temperature is quite low, such as $T_C = -100°C$, the corresponding Fahrenheit temperature is found to be $T_F = \dfrac{9}{5}(-100) + 32 = -148°F$ or the Fahrenheit scale reading is lower than the corresponding reading on the Celsius scale. Thus, we see that the graphs of the readings on these two temperature scales must cross one another, and there must be some temperature at which the two temperature scales will read the same.

To find the temperature which has the same numeric value on both the Celsius and the Fahrenheit scale, we substitute $T_C = T_F$ into the conversion equation. This gives

$$T_F = \frac{9}{5} T_F + 32 \quad \text{or} \quad -\frac{4}{5} T_F = 32$$

and yields $T_F = -\dfrac{5}{4}(32) = -40°$ as the single temperature at which the readings on the two scales have the same numeric value.   ◊

15.   A brass ring of diameter 10.00 cm at 20.0 °C is heated and slipped over an aluminum rod of diameter 10.01 cm at 20.0 °C. Assuming the average coefficients of linear expansion are constant, (a) to what temperature must the combination be cooled to separate the two metals? Is that temperature attainable? (b) What if the aluminum rod were 10.02 cm in diameter?

Solution

The magnitude, L, of a linear dimension on an object at temperature T is given by

$$L = L_0\left[1+\alpha(T-T_0)\right] = L_0 + \alpha L_0(T-T_0)$$

where L_0 is the magnitude of that dimension at temperature T_0 and α is the coefficient of linear expansion of the material making up the object.

To remove the ring from the rod, the diameter of the ring must be at least as large as the diameter of the rod.

Thus, we require that $\qquad\left(L_f\right)_{\text{brass}} = \left(L_f\right)_{\text{Al}}$

or $\qquad\qquad L_{\text{brass}} + \alpha_{\text{brass}} L_{\text{brass}}(\Delta T) = L_{\text{Al}} + \alpha_{\text{Al}} L_{\text{Al}}(\Delta T)$

This gives $\qquad\qquad \Delta T = \dfrac{L_{\text{Al}} - L_{\text{brass}}}{\alpha_{\text{brass}} L_{\text{brass}} - \alpha_{\text{Al}} L_{\text{Al}}}$

(a)   If $L_{\text{Al}} = 10.01$ cm and $L_{\text{brass}} = 10.00$ cm at $T = 20.0$ °C, the change in temperature required to make the new lengths equal is

$$\Delta T = \frac{10.01 - 10.00}{\left[19 \times 10^{-6}\ (\text{°C})^{-1}\right](10.00) - \left[24 \times 10^{-6}\ (\text{°C})^{-1}\right](10.01)} = -199\ \text{°C}$$

The final temperature will then be

$\qquad T_f = T_0 + \Delta T = 20.0\ \text{°C} - 199\ \text{°C} = -179\ \text{°C}$ which is attainable.   ◊

(b)   If $L_{\text{Al}} = 10.02$ cm   and $L_{\text{brass}} = 10.00$ cm   at $T = 20.0$ °C,   the change in temperature required to make the new lengths equal is

$$\Delta T = \frac{10.02 - 10.00}{\left[19 \times 10^{-6}\ (\text{°C})^{-1}\right](10.00) - \left[24 \times 10^{-6}\ (\text{°C})^{-1}\right](10.02)} = -396\ \text{°C}$$

The final temperature will then be

$\qquad T_f = T_0 + \Delta T = 20.0\ \text{°C} - 396\ \text{°C} = -376\ \text{°C}$

which is below absolute zero $(-273.15\ \text{°C})$ and, therefore, unattainable.   ◊

21.   An automobile fuel tank is filled to the brim with 45 L (12 gal) of gasoline at 10°C. Immediately afterward, the vehicle is parked in the sunlight, where the temperature is 35°C. How much gasoline overflows from the tank as a result of the expansion? (Neglect the expansion of the tank.)

Solution

If a solid or liquid occupies volume V_0 at temperature T_0, its volume at temperature T will be given by

$$V = V_0\left[1+\beta\left(T-T_0\right)\right] = V_0 + \beta V_0\left(T-T_0\right)$$

where β is the coefficient of volume expansion of that material. Thus, the change in volume that occurs is

$$\Delta V = V - V_0 = \beta V_0\left(T-T_0\right)$$

The coefficient of volume expansion for gasoline is $\beta = 9.6\times10^{-4}\ (°C)^{-1}$. Therefore, if the 45-L fuel tank is filled to the brim with gasoline at $T_0 = 10\ °C$, the increase in volume (and hence the amount of overflow) as the temperature rises to $T = 35\ °C$ will be

$$\Delta V = \left[9.6\times10^{-4}\ (°C)^{-1}\right](45\ L)(35\ °C-10\ °C) = 1.1\ L \qquad \Diamond$$

27.   One mole of oxygen gas is at a pressure of 6.00 atm and a temperature of 27.0°C. (a) If the gas is heated at constant volume until the pressure triples, what is the final temperature? (b) If the gas is heated so that both the pressure and volume are doubled, what is the final temperature?

Solution

(a)   We shall treat the oxygen as an ideal gas and apply the ideal gas law, $PV = nRT$. When n and V are both held constant, we see that the ratio P/T will be constant. Thus,

$$\frac{P_f}{T_f} = \frac{P_i}{T_i} \qquad \text{or} \qquad T_f = T_i\left(\frac{P_f}{P_i}\right)$$

Remembering to always use absolute temperatures with the ideal gas law $(T_i = 27.0 + 273.15 = 300 \text{ K})$, this yields

$$T_f = (300 \text{ K})\left(\frac{3P_i}{P_i}\right) = 900 \text{ K} \qquad \text{or} \qquad (T_C)_f = 900 - 273.15 = 627°C \qquad \Diamond$$

(b)   Again using the ideal gas law, we write $T = \dfrac{PV}{nR}$ for both the initial and final states, obtaining

$$\frac{T_f}{T_i} = \frac{(P_f V_f / nR)}{(P_i V_i / nR)} = \frac{P_f V_f}{P_i V_i} = \frac{(2P_i)(2V_i)}{P_i V_i} = 4$$

Thus, $T_f = 4T_i = 4(300 \text{ K}) = 1\,200 \text{ K}$ or $\quad (T_C)_f = 1\,200 - 273.15 = 927°C \qquad \Diamond$

33.   A weather balloon is designed to expand to a maximum radius of 20 m at its working altitude, where the air pressure is 0.030 atm and the temperature is 200 K. If the balloon is filled at atmospheric pressure and 300 K, what is its radius at liftoff?

Solution

We neglect any effects due to elastic forces in the material making up the skin of the balloon, and assume that the gas pressure inside the balloon always equals the atmospheric pressure outside the balloon. Then, the thermodynamic properties of the gas at liftoff are:

$$P_i = 1.0 \text{ atm}, \ T_i = 300 \text{ K}, \text{ and } V_i = 4\pi r_i^3 / 3$$

The properties of the gas when the balloon is at the working altitude are:

$$P_f = 0.030 \text{ atm}, \ T_f = 200 \text{ K}, \text{ and } V_f = 4\pi r_f^3 / 3 = 4\pi (20 \text{ m})^3 / 3$$

Using the ideal gas law and recognizing that $n_f = n_i$ (assuming no leakage as the balloon ascends), we obtain

$$V_i = V_f \left(\frac{P_f}{P_i} \right) \left(\frac{T_i}{T_f} \right) \quad \text{or} \quad \left(\frac{4\pi}{3} \right) r_i^3 = \left(\frac{4\pi}{3} \right) (20 \text{ m})^3 \left(\frac{P_f}{P_i} \right) \left(\frac{T_i}{T_f} \right)$$

Thus, $r_i = (20 \text{ m}) \left[\left(\frac{0.030 \text{ atm}}{1.0 \text{ atm}} \right) \left(\frac{300 \text{ K}}{200 \text{ K}} \right) \right]^{\frac{1}{3}} = 7.1 \text{ m}$ ◊

41.   A cylinder contains a mixture of helium and argon gas in equilibrium at a temperature of 150 °C. (a) What is the average kinetic energy of each type of molecule? (b) What is the rms speed of each type of molecule?

Solution

(a)   The absolute temperatures of both the helium and argon gases are

$$T = T_C + 273.15 = 150 + 273.15 = 423 \text{ K}$$

Therefore, the average kinetic energy of a molecule in either gas is given by

$$\overline{KE}_{\text{molecule}} = \tfrac{3}{2}k_B T_K = \tfrac{3}{2}(1.38\times10^{-23} \text{ J/K})(423 \text{ K}) = 8.76\times10^{-21} \text{ J} \qquad \diamond$$

(b)   Since $\overline{KE}_{\text{molecule}} = \tfrac{1}{2}mv_{rms}^2$, where m is the mass of the molecule and v_{rms} is the root-mean-square speed of the molecules in that gas,

we have $\qquad\qquad\qquad v_{rms} = \sqrt{\dfrac{2\overline{KE}_{\text{molecule}}}{m}}$

For helium, $m = \dfrac{M}{N_A} = \dfrac{4.00\times10^{-3} \text{ kg/mol}}{6.02\times10^{23} \text{ molecules/mol}} = 6.64\times10^{-27} \text{ kg}$

and $\qquad\qquad v_{rms} = \sqrt{\dfrac{2(8.76\times10^{-21} \text{ J})}{6.64\times10^{-27} \text{ kg}}} = 1.62\times10^3 \text{ m/s} = 1.62 \text{ km/s} \qquad \diamond$

For argon, $m = \dfrac{M}{N_A} = \dfrac{39.9\times10^{-3} \text{ kg/mol}}{6.02\times10^{23} \text{ molecules/mol}} = 6.63\times10^{-26} \text{ kg}$

and $\qquad\qquad v_{rms} = \sqrt{\dfrac{2(8.76\times10^{-21} \text{ J})}{6.63\times10^{-26} \text{ kg}}} = 514 \text{ m/s} \qquad \diamond$

43. Superman leaps in front of Lois Lane to save her from a volley of bullets. In a 1-min interval, an automatic weapon fires 150 bullets, each of mass 8.0 g, at 400 m/s. The bullets strike his mighty chest, which has an area of 0.75 m². Find the average force exerted on Superman's chest if the bullets bounce back after an elastic, head-on collision.

Solution

Choosing the positive direction to be directed from the shooter toward Superman, the initial velocity of a bullet is $\vec{v}_i = +400$ m/s. Since the bullets undergo an elastic head-on collision, their final velocity is $\vec{v}_f = -400$ m/s. The impulse experienced by each bullet is

$$\text{Impulse})_{\text{bullet}} = \vec{I} = m\vec{v}_f - m\vec{v}_i$$

$$= \left(8.0 \times 10^{-3} \text{ kg}\right)\left[-400 \text{ m/s} - \left(+400 \text{ m/s}\right)\right] = -6.4 \text{ kg} \cdot \text{m/s}$$

From Newton's third law, each bullet then imparts an impulse of $+6.4$ kg·m/s to Superman's chest. The total impulse he experiences in a time interval of $\Delta t = 1.0$ min $= 60$ s is given by

$$I_{total} = 150\left(+6.4 \text{ kg} \cdot \text{m/s}\right) = +9.6 \times 10^2 \text{ kg} \cdot \text{m/s}$$

and the average force exerted on his chest is

$$\vec{F}_{av} = \frac{\vec{I}_{total}}{\Delta t} = \frac{+9.6 \times 10^2 \text{ kg} \cdot \text{m/s}}{60 \text{ s}} = +16 \text{ N}$$

or $\quad \vec{F}_{av} = 16$ N directed from the shooter toward Superman. $\qquad \Diamond$

47. A popular brand of cola contains 6.50 g of carbon dioxide dissolved in 1.00 L of soft drink. If the evaporating carbon dioxide is trapped in a cylinder at 1.00 atm and 20.0°C, what volume does the gas occupy?

Solution

The molecular weight of carbon dioxide (CO_2) is

$$M_{CO_2} = (\text{atomic mass of carbon}) + 2(\text{atomic mass of oxygen})$$
$$= 12.0 + 2(16.0) = 44.0 \text{ g/mol}$$

The number of moles of carbon dioxide dissolved in the 1.00-L of soft drink is

$$n = \frac{m}{M_{CO_2}} = \frac{6.50 \text{ g}}{44.0 \text{ g/mol}} = 0.148 \text{ mol}$$

Under the specified conditions,

$$P = 1.00 \text{ atm} = 1.013 \times 10^5 \text{ Pa}, \ T = 20°C = 293 \text{ K}$$

the ideal gas law gives the volume occupied by this quantity of gas as

$$V = \frac{nRT}{P} = \frac{(0.148 \text{ mol})(8.31 \text{ J/mol} \cdot \text{K})(293 \text{ K})}{1.013 \times 10^5 \text{ Pa}}$$

or $V = 3.55 \times 10^{-3} \text{ m}^3 = 3.55 \text{ L}$ ◊

51. A liquid with coefficient of volume expansion of β just fills a spherical flask of volume V_0 at temperature T (Fig. P10.51). The flask is made of a material that has a coefficient of linear expansion of α. The liquid is free to expand into a capillary of cross-sectional area A at the top. (a) Show that if the temperature increases by ΔT, the liquid rises in the capillary by the amount $\Delta h = (V_0/A)(\beta - 3\alpha)\Delta T$. (b) For a typical system, such as a mercury thermometer, why is it a good approximation to neglect the expansion of the flask?

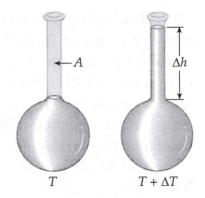

Figure P10.51

Solution

(a) When the temperature increases by ΔT, the volume of the liquid expands by an amount

$$\Delta V_{\text{liquid}} = \beta V_0 (\Delta T)$$

and the volume enclosed by the spherical flask increases by the amount

$$\Delta V_{\text{flask}} = (3\alpha) V_0 (\Delta T)$$

The volume of liquid which must overflow the flask and rise into the capillary is

$$V_{\text{overflow}} = \Delta V_{\text{liquid}} - \Delta V_{\text{flask}} = A(\Delta h)$$

where A is the cross-sectional area of the capillary and Δh is the height to which the liquid will rise in the capillary.

Therefore, $$\Delta h = \frac{V_{\text{overflow}}}{A} = \frac{\beta V_0 (\Delta T) - (3\alpha) V_0 (\Delta T)}{A}$$

$$\Delta h = \left(\frac{V_0}{A}\right)(\beta - 3\alpha)\Delta T \qquad\qquad \Diamond$$

(b) For a mercury thermometer, $\beta_{\text{Hg}} = 1.82 \times 10^{-4} \ (°\text{C})^{-1}$

and (assuming Pyrex glass), $3\alpha_{\text{glass}} = 3(3.2 \times 10^{-6} \ (°\text{C})^{-1}) = 9.6 \times 10^{-6} \ (°\text{C})^{-1}$

Thus, the expansion of the mercury is almost 20 times that of the glass flask. This means that it is a good approximation to neglect the expansion of the flask. $\qquad \Diamond$

54.   Before beginning a long trip on a hot day, a driver inflates an automobile tire to a gauge pressure of 1.80 atm at 300 K. At the end of the trip the gauge pressure has increased to 2.20 atm. (a) Assuming that the volume has remained constant, what is the temperature of the air inside the tire? (b) What percentage of the original mass of air in the tire should be released so the pressure returns to its original value? Assume that the temperature remains at the value found in (a) and the volume of the tire remains constant as air is released.

Solution

(a)   It is important to remember that one must always use the absolute pressure and the absolute temperature when applying the ideal gas law. The initial and final absolute pressures of the air in the tire are

$$P_i = P_{atm} + \left(P_{gauge}\right)_i = 1.00 \text{ atm} + 1.80 \text{ atm} = 2.80 \text{ atm}$$

and $$P_f = P_{atm} + \left(P_{gauge}\right)_f = 1.00 \text{ atm} + 2.20 \text{ atm} = 3.20 \text{ atm}$$

Thus, with volume constant, the ideal gas law gives

$$T_f = T_i \left(\frac{P_f}{P_i}\right) = (300 \text{ K})\left(\frac{3.20 \text{ atm}}{2.80 \text{ atm}}\right) = 343 \text{ K} \qquad \Diamond$$

(b)   From the ideal gas law, $PV = nRT$, the number of moles of gas present is $n = PV/RT$. Thus, if the volume enclosed by the tire and the temperature of the gas in the tire remain constant as gas is released, we have

$$\frac{n_f}{n_i} = \frac{P_f V/RT}{P_i V/RT} = \frac{P_f}{P_i} = \frac{2.80 \text{ atm}}{3.20 \text{ atm}} = 0.875$$

or 87.5% of the number of moles originally in the tire is still present after the pressure is adjusted. Thus, 12.5% of the gas originally in the tire was released.

Since the mass of the gas is directly proportional to the number of moles present, 12.5% of the original mass of air in the tire was released. $\Diamond$

61.   A bimetallic bar is made of two thin strips of dissimilar metals bonded together. As they are heated, the one with the larger average coefficient of expansion expands more than the other, forcing the bar into an arc, with the outer strip having both a larger radius and a larger circumference. (See Fig. P10.61.) (a) Derive an expression for the angle of bending, θ, as a function of the initial length of the strips, their average coefficients of linear expansion, the change in temperature, and the separation of the centers of the strips ($\Delta r = r_2 - r_1$). (b) Show that the angle of bending goes to zero when ΔT goes to zero or when the two coefficients of expansion become equal. (c) What happens if the bar is cooled?

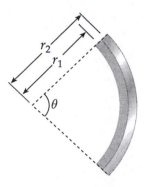

Figure P10.61

Solution

(a)   Let L_i be the common initial length of the two strips. After a change in temperature ΔT, their lengths are

$$L_1 = L_i + \alpha_1 L_i (\Delta T) \qquad \text{and} \qquad L_2 = L_i + \alpha_2 L_i (\Delta T)$$

The lengths of the circular arcs are related to their radii by $L_1 = r_1 \theta$ and $L_2 = r_2 \theta$, where θ is measured in radians. Thus,

$$\Delta r = r_2 - r_1 = \frac{L_2}{\theta} - \frac{L_1}{\theta} = \frac{(\alpha_2 - \alpha_1) L_i (\Delta T)}{\theta} \qquad \text{or} \qquad \theta = \frac{(\alpha_2 - \alpha_1) L_i (\Delta T)}{\Delta r} \qquad \Diamond$$

(b)   As seen in the above result, $\theta = 0$ if either $\Delta T = 0$ or $\alpha_1 = \alpha_2$. $\Diamond$

(c)   If $\Delta T < 0$, then θ is negative, meaning the bar bends in the direction opposite to that shown. $\Diamond$

Chapter 11

Energy in Thermal Processes

NOTES FROM SELECTED CHAPTER SECTIONS

11.1 Heat and Internal Energy

When two systems at different temperatures are placed in thermal contact, energy is transferred by heat from the warmer to the cooler object until they reach a common temperature (that is, when they are in thermal equilibrium with each other). Heat is a mechanism by which energy (measured in calories or joules) is transferred between a system and its environment due to a temperature difference.

The **mechanical equivalent of heat,** first measured by Joule, is given by $1 \, cal = 4.186 \, J$. This is the definition of the calorie.

11.2 Specific Heat

The **specific heat,** c, of any substance is defined as the amount of energy required to increase the temperature of a unit mass of that substance by one Celsius degree. The units of specific heat are $J/kg \cdot {}^\circ C$.

11.3 Calorimetry

Calorimetry is a procedure carried out in an isolated system (usually a substance of unknown specific heat and water) in which the transfer of thermal energy occurs. A negligible amount of mechanical work is done in the process. The law of conservation of energy in a calorimeter requires that the energy that leaves the warmer substance (of unknown specific heat) equals the energy that enters the water.

11.4 Latent Heat and Phase Change

A substance usually undergoes a change in temperature when energy is transferred by heat between it and its surroundings. There are situations, however, in which the transfer of energy does not result in a change in temperature. This is the case whenever the substance undergoes a physical alteration from one form to another, referred to as a **phase change**. Some common phase changes are solid to liquid, liquid to gas, and a change in crystalline structure of a solid. Every phase change involves a change in internal energy without an accompanying change in temperature.

The **latent heat of fusion** is a parameter used to characterize a solid-to-liquid phase change; the **latent heat of vaporization** characterizes the liquid-to-gas phase change.

11.5 Energy Transfer

There are three basic processes of thermal energy transfer: conduction, convection, and radiation.

Conduction is an energy transfer process which occurs in a substance when there is a temperature gradient across the substance. That is, conduction of energy occurs only when the temperature of the substance is **not** uniform. For example, if you position a metal rod with one end in a flame, energy will flow from the hot end to the colder end. The rate of flow of heat along the rod is proportional to the cross-sectional area of the rod, the temperature gradient, and k, the thermal conductivity of the material of which the rod is made.

Convection is a process of energy transfer by the motion of material, such as the mixing of hot and cold fluids. Convection heating is used in conventional hot-air and hot-water heating systems. Convection currents produce changes in weather conditions when warm and cold air masses mix in the atmosphere.

Radiation is energy transfer by the emission of electromagnetic waves.

EQUATIONS AND CONCEPTS

When two systems initially at different temperatures are placed in contact with each other, energy will be transferred from the system at higher temperature to the system at lower temperature until the two systems reach a common temperature (thermal equilibrium).

Thermal equilibrium

The **mechanical equivalent of heat** was first measured by Joule. This is the definition of the calorie as an energy unit.

$$1 \text{ cal} = 4.186 \text{ J} \tag{11.1}$$

The quantity of energy required to increase the temperature of a given mass by a specified amount varies from one substance to another. Every substance is characterized by a unique value of **specific heat**, c. *In this and subsequent equations, Q represents the quantity of energy transferred by heat between a system and its environment.*

$$Q = mc\,\Delta T \tag{11.3}$$

The energy required to cause a quantity of substance of mass, *m*, to undergo a phase change depends on the value of the **latent heat** of the substance. The latent heat of fusion, L_f is used when the phase change is from solid to liquid (or liquid to solid). The latent heat of vaporization, L_v is used when the phase change is from liquid to gas (or gas to liquid). The positive or negative sign is chosen according to the direction of energy flow.

$$Q = \pm mL \tag{11.6}$$

This is the basic **law of thermal conduction.** The constant k is called the thermal conductivity and is characteristic of a particular material. Conduction of energy occurs only when there is a temperature gradient (or temperature difference between two points in the conducting material).

$$P = kA\left(\frac{T_h - T_c}{L}\right) \tag{11.7}$$

To calculate the rate of energy transfer through a **compound slab** a summation is made over all portions of the slab. For this calculation, T_h and T_c are the temperatures of the outer extremities of the slab. In engineering practice, the term L/k for a particular substance is referred to as the **R value** of the material.

$$\frac{Q}{\Delta t} = \frac{A(T_h - T_c)}{\sum_i L_i/k_i} \tag{11.8}$$

The **rate of energy transfer,** P, can be expressed in cal/s, Btu/h, or watts (where $1\,W = 1\,J/s$ and $1\,Btu/h = 0.293\,W$).

Convection is a process in which energy transfer occurs as the result of the movement of a substance (usually a fluid) between points at different temperatures.

Stefan's law expresses the rate of emission of heat energy by **radiation** (the power radiated). *The radiated power is proportional to the fourth power of the absolute temperature.*

$$P = \sigma A e T^4 \tag{11.10}$$

$\sigma = 5.669 \times 10^{-8}\ W/m^2$

$T =$ temperature in kelvins

$e =$ emissivity (values from 0 to 1)

An object at temperature T in surroundings at temperature T_0 will experience a net radiated ($T > T_0$) or absorbed ($T < T_0$) power. *At thermal equilibrium* ($T = T_0$), *an object radiates and absorbs energy at the same rate and the temperature of the object remains constant.*

$$P_{net} = \sigma A e\left(T^4 - T_0^4\right) \tag{11.11}$$

SUGGESTIONS, SKILLS, AND STRATEGIES

If you are having difficulty with calorimetry problems, one or more of the following factors should be considered:

1. Be sure your units are consistent throughout. That is, if you are using specific heats measured in $J/kg \cdot °C$, be sure that masses are in kilograms and temperatures are in Celsius units throughout.

2. Losses and gains in energy are found by using $Q = mc\Delta T$ only for those intervals in which no phase changes occur. Likewise, the equations $Q = \pm mL_f$ and $Q = \pm mL_v$ are to be used only when phase changes are taking place.

3. Often sign errors occur applying the basic calorimetry equation ($Q_{cold} = -Q_{hot}$). Remember to include the negative sign in the equation; and remember that ΔT **is always the final temperature minus the initial temperature**.

REVIEW CHECKLIST

▷ Understand the important distinction between internal energy and heat.

▷ Define and discuss the calorie, Btu, specific heat, and latent heat. Convert among calories, Btu's, and joules.

▷ Use equations for specific heat, latent heat, temperature change and energy gain (loss) to solve calorimetry problems.

▷ Discuss the possible mechanisms which can give rise to energy transfer between a system and its surroundings; that is, conduction, convection and radiation; and give a realistic example of each transfer mechanism.

▷ Apply the basic law of thermal conduction, and Stefan's law for energy transfer by radiation.

SOLUTIONS TO SELECTED END-OF-CHAPTER PROBLEMS

3.   Lake Erie contains roughly 4.00×10^{11} m^3 of water. (a) How much energy is required to raise the temperature of that volume of water from 11.0°C to 12.0°C? (b) How many years would it take to supply this amount of energy by using the 1 000-MW exhaust energy of an electric power plant?

Solution

(a)   The mass of the water in Lake Erie is

$$m = \rho V = \left(1.00 \times 10^3 \text{ kg/m}^3\right)\left(4.00 \times 10^{11} \text{ m}^3\right) = 4.00 \times 10^{14} \text{ kg}$$

and the energy needed to raise the temperature of this water from 11.0°C to 12.0°C is given by $Q = mc(\Delta T)$ as

$$Q = \left(4.00 \times 10^{14} \text{ kg}\right)\left(4\,186 \text{ J/kg} \cdot {}^\circ\text{C}\right)\left(12.0^\circ\text{C} - 11.0^\circ\text{C}\right) = 1.67 \times 10^{18} \text{ J} \qquad \lozenge$$

(b)   The exhaust from the power plant would add energy to the water of Lake Erie at a rate of

$$\mathcal{P} = 1\,000 \text{ MW} = 1\,000 \times 10^6 \text{ W} = 1.00 \times 10^9 \text{ J/s}$$

and the time needed to raise the temperature of the lake from 11.0°C to 12.0°C is

$$t = \frac{Q}{\mathcal{P}} = \frac{1.67 \times 10^{18} \text{ J}}{1.00 \times 10^9 \text{ J/s}} = 1.67 \times 10^9 \text{ s}\left(\frac{1 \text{ yr}}{3.156 \times 10^7 \text{ s}}\right) = 53.1 \text{ yr} \qquad \lozenge$$

7.   A 75.0-kg weight watcher wishes to climb a mountain to work off the equivalent of a large piece of chocolate cake rated at 500 (food) Calories. How high must the person climb? [1 (food) Calorie $= 10^3$ calories]

Solution

The chemical energy which must be converted into other forms during the climb is

$$Q = 500 \text{ Calories}\left(\frac{10^3 \text{ cal}}{1 \text{ Calorie}}\right) = 5.00 \times 10^5 \text{ cal}$$

Using the mechanical equivalent of heat to convert this to standard SI units of energy gives

$$Q = 5.00 \times 10^5 \text{ cal}\left(\frac{4.186 \text{ J}}{1 \text{ cal}}\right) = 2.09 \times 10^6 \text{ J}$$

If we assume that this all goes to increasing gravitational potential energy as the person climbs, then $Q = \Delta PE_g = mgh$. The required height of the climb is then

$$h = \frac{Q}{mg} = \frac{2.09 \times 10^6 \text{ J}}{(75.0 \text{ kg})(9.80 \text{ m/s}^2)} = 2.85 \times 10^3 \text{ m} = 2.85 \text{ km} \qquad \Diamond$$

In actual practice, much of the chemical energy would be dissipated by radiation and by evaporation of perspiration as the person climbs, so the required height would be considerably smaller than that calculated above.

15. An aluminum cup contains 225 g of water and a 40-g copper stirrer all at 27 °C. A 400-g sample of silver at an initial temperature of 87 °C is placed in the water. The stirrer is used to stir the mixture until it reaches its final equilibrium temperature of 32 °C. Calculate the mass of the aluminum cup.

Solution

Assuming that the system consisting of the cup, stirrer, water, and silver sample is isolated from the environment, the total energy content of the system is constant. Therefore, any energy gained by the initially cooler parts (cup, stirrer, and water) must equal the energy lost by the initially warmer parts (the silver sample).

Thus, $$Q_{cup} + Q_{stirrer} + Q_{water} = -Q_{Ag}$$

Since the cup, stirrer, and water all undergo the same change in temperature, this becomes

$$\left[m_{cup}c_{Al} + m_{stirrer}c_{Cu} + m_{water}c_{water} \right] \left(\Delta T_{water} \right) = -m_{Ag}c_{Ag} \left(\Delta T_{Ag} \right)$$

The mass of the cup is then

$$m_{cup} = m_{Ag} \left(\frac{c_{Ag}}{c_{Al}} \right) \left[\frac{-\Delta T_{Ag}}{\Delta T_{water}} \right] - m_{stirrer} \left(\frac{c_{Cu}}{c_{Al}} \right) - m_{water} \left(\frac{c_{water}}{c_{Al}} \right)$$

or $$m_{cup} = \left(400 \text{ g} \right) \left(\frac{234}{900} \right) \left[\frac{-(32-87)}{32-27} \right] - \left(40 \text{ g} \right) \left(\frac{387}{900} \right) - \left(225 \text{ g} \right) \left(\frac{4\,186}{900} \right) = 80 \text{ g} \quad \Diamond$$

19. A student drops two metallic objects into a 120-g steel container holding 150 g of water at 25 °C. One object is a 200-g cube of copper that is initially at 85 °C, and the other is a chunk of aluminum that is initially at 5.0 °C. To the surprise of the student, the water reaches a final temperature of 25 °C, precisely where it started. What is the mass of the aluminum chunk?

Solution

Notice that the initial temperature of the copper cube is higher than the final temperature. Thus, the copper cube loses energy by heat in this process. The initial temperature of the aluminum chunk is less than the final temperature, so it gains internal energy. The temperature of the water and the steel container is unchanged so these parts of the system have zero net change in internal energy. Thus, assuming the system is isolated from its surroundings, the energy balance equation is

$$Q_{Al} = -Q_{Cu} \qquad \text{or} \qquad m_{Al} c_{Al} (\Delta T_{Al}) = -m_{Cu} c_{Cu} (\Delta T_{Cu})$$

The specific heats of aluminum and copper are

$$c_{Al} = 900 \text{ J/kg} \cdot {}^{\circ}\text{C} \qquad \text{and} \qquad c_{Cu} = 387 \text{ J/kg} \cdot {}^{\circ}\text{C}$$

so the mass of the aluminum chunk is

$$m_{Al} = m_{Cu} \left(\frac{c_{Cu}}{c_{Al}} \right) \left[\frac{-(\Delta T_{Cu})}{\Delta T_{Al}} \right] = (200 \text{ g}) \left(\frac{387}{900} \right) \left[\frac{-(25-85)}{25-5.0} \right] = 260 \text{ g} = 0.26 \text{ kg} \qquad \Diamond$$

25.   A 75-kg cross-country skier glides over snow as in Figure P11.25. The coefficient of friction between skis and snow is 0.20. Assume all the snow beneath his skis is at 0°C and that all the internal energy generated by friction is added to snow, which sticks to his skis until it melts. How far would he have to ski to melt 1.0 kg of snow?

Solution

The snow is already at the melting temperature of 0°C, so the energy that must be added to convert a kilogram of snow into liquid water at 0°C is

$$Q = mL_f = (1.0\ \text{kg})(3.33 \times 10^5\ \text{J/kg}) = 3.33 \times 10^5\ \text{J}$$

Since the skier is on level ground, the normal force exerted on the skis by the snow equals the skier's weight, $n = mg$

and the friction force is $f_k = \mu_k n = \mu_k mg$

The magnitude of the work done by the friction force when the skier goes a distance x is $|W_f| = f \cdot x$. If all of the work done by the friction is added to the snow as internal energy, the distance the skier must travel to melt 1.0 kg of snow is given by

$$|W_f| = f \cdot x = Q$$

or $$x = \frac{Q}{f} = \frac{Q}{\mu_k mg} = \frac{3.33 \times 10^5\ \text{J}}{(0.20)(75\ \text{kg})(9.8\ \text{m/s}^2)} = 2.3 \times 10^3\ \text{m} = 2.3\ \text{km} \qquad \Diamond$$

31.   Steam at 100°C is added to ice at 0°C. (a) Find the amount of ice melted and the final temperature when the mass of steam is 10 g and the mass of ice is 50 g. (b) Repeat with steam of mass 1.0 g and ice of mass 50 g.

Solution

(a)   With a fairly high ratio of steam to ice, we shall assume that all the ice will melt and the equilibrium temperature is somewhere above 0°C.

Equating the energy gained by the ice to the energy given up by the steam gives

$$m_{ice}L_f + m_{ice}c_{water}(\Delta T)_{melted \atop ice} = m_{steam}L_v + m_{steam}c_{water}\left[-(\Delta T)_{condensed \atop steam}\right]$$

or

$$(50 \text{ g})(79.7 \text{ cal/g}) + (50 \text{ g})(1.0 \text{ cal/g} \cdot °C) (T_f - 0°C) =$$
$$(10 \text{ g})(540 \text{ cal/g}) + (10 \text{ g})(1.0 \text{ cal/g} \cdot °C)\left[-(T_f - 100°C)\right]$$

Solving for the final temperature gives

$$T_f = \frac{5\,400 \text{ cal} + 1\,000 \text{ cal} - 4\,000 \text{ cal}}{50 \text{ cal/°C} + 10 \text{ cal/°C}} = 40°C$$ ◊

(b)   If we attempt to duplicate the previous calculation when only 1.0 g of steam is used, we obtain

$$(50 \text{ g})(79.7 \text{ cal/g}) + (50 \text{ g})(1.0 \text{ cal/g} \cdot °C) (T_f - 0°C) =$$
$$(1.0 \text{ g})(540 \text{ cal/g}) + (1.0 \text{ g})(1.0 \text{ cal/g} \cdot °C)\left[-(T_f - 100°C)\right]$$

and $$T_f = \frac{540 \text{ cal} + 100 \text{ cal} - 4\,000 \text{ cal}}{50 \text{ cal/°C} + 1.0 \text{ cal/°C}} = -66°C$$

which is clearly incorrect. Adding steam to ice will not lower the temperature of the ice!

Carefully analyzing the previous solution shows that we assumed all the ice will melt. However, if only 1.0 g of steam is used, the total energy the steam can give up as it condenses and cools all the way to 0°C (the temperature of the ice) is

$$Q' = m_{steam}L_v + m_{steam}c_{water}\left[-(\Delta T)_{\substack{condensed \\ steam}}\right]$$

$$= (1.0 \text{ g})(540 \text{ cal/g}) + (1.0 \text{ g})(1.0 \text{ cal/g} \cdot °C)(100°C) = 640 \text{ cal}$$

The amount of ice this much energy can melt is only

$$m = \frac{Q'}{L_f} = \frac{640 \text{ cal}}{79.7 \text{ cal/g}} = 8.0 \text{ g}$$

Thus, we see that 42 grams of ice will remain and that all parts of the system (the remaining ice, the melted ice, and the condensed steam) come to an equilibrium temperature of 0°C in this case.   ◊

35.   A steam pipe is covered with 1.50-cm-thick insulating material of thermal conductivity 0.200 cal/cm·°C·s. How much energy is lost every second when the steam is at 200°C and the surrounding air is at 20.0°C? The pipe has a circumference of 800 cm and a length of 50.0 m. Neglect losses through the ends of the pipe.

Solution

The area of the inner surface of the insulation layer covering the cylindrical pipe is

$$A = (circumference) \times (length) = (800 \text{ cm}) \times (50.0 \times 10^2 \text{ cm}) = 4.00 \times 10^6 \text{ cm}^2$$

The rate of energy transfer by conduction through the insulation layer is given by the relation

$$\mathscr{P} = kA \left(\frac{T_h - T_c}{L} \right)$$

where k is the thermal conductivity of the material, A is the surface area of the conducting material, L is the thickness of the material, and $\Delta T = T_h - T_c$ is the difference in temperature on the two sides of the conducting material.

Thus, the rate of energy loss through the side of the steam pipe is

$$\mathscr{P} = \left(0.200 \ \frac{\text{cal}}{\text{cm} \cdot \text{°C} \cdot \text{s}} \right) \left(4.00 \times 10^6 \text{ cm}^2 \right) \left(\frac{200°C - 20.0°C}{1.50 \text{ cm}} \right) = 9.60 \times 10^7 \ \frac{\text{cal}}{\text{s}}$$

or $\mathscr{P} = 9.60 \times 10^7 \ \dfrac{\text{cal}}{\text{s}} \left(\dfrac{4.186 \text{ J}}{1 \text{ cal}} \right) = 4.02 \times 10^8 \text{ J/s} = 402 \text{ MW}$ ◊

Clearly, this pipe needs to be better insulated.

---

39.   A copper rod and an aluminum rod of equal diameter are joined end to end in good thermal contact. The temperature of the free end of the copper rod is held constant at 100°C, and that of the far end of the aluminum rod is held at 0°C. If the copper rod is 0.15 m long, what must be the length of the aluminum rod so that the temperature at the junction is 50°C?

Solution

The rate at which a rod of thermal conductivity k, length L, and cross-sectional area A will transfer energy by conduction is

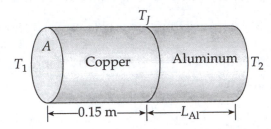

$$\mathcal{P} = kA\left(\frac{T_h - T_c}{L}\right)$$

Here, T_h is the temperature of the hotter end of the rod while T_c is the temperature of the cooler end.

When the temperature of the junction between the cooper rod and the aluminum rod has stabilized at $T_J = 50°C$, the copper rod must be transferring energy up to the junction at the same rate as the aluminum rod is transferring energy away from the junction. Hence,

$$k_{Cu} A\left(\frac{T_1 - T_J}{L_{Cu}}\right) = k_{Al} A\left(\frac{T_J - T_2}{L_{Al}}\right)$$

or the length of the aluminum rod must be

$$L_{Al} = L_{Cu}\left(\frac{k_{Al}}{k_{Cu}}\right)\left(\frac{T_J - T_2}{T_1 - T_J}\right) = (0.15 \text{ m})\left(\frac{238 \text{ J/s} \cdot \text{m} \cdot °C}{397 \text{ J/s} \cdot \text{m} \cdot °C}\right)\left(\frac{50°C - 0°C}{100°C - 50°C}\right)$$

or $L_{Cu} = 0.090 \text{ m} = 9.0 \text{ cm}$ ◊

46.   At high noon, the Sun delivers 1.00 kW to each square meter of a blacktop road. If the hot asphalt loses energy only by radiation, what is its equilibrium temperature?

Solution

When the asphalt has reached equilibrium temperature, energy must be transferred away from the asphalt at the same rate as the Sun delivers energy to the asphalt.

If energy transfers from the asphalt by conduction and convection may be neglected, then the rate at which the asphalt loses energy by radiation must match the power input from the Sun, or

$$\mathcal{P}_{radiation} = \sigma A e T^4 = \mathcal{P}_{\substack{input \\ from\ Sun}} = A(1.00\ \text{kW/m}^2) = A(1.00\times10^3\ \text{W/m}^2)$$

Here, $\sigma = 5.67\times10^{-8}\ \text{W/m}^2\cdot\text{K}^4$, A is the area of the radiating surface, e is the emissivity of the surface, and T is the absolute temperature of the radiator.

In order to obtain the minimum possible temperature of the asphalt, we approximate it as a perfect radiator with emissivity $e = 1.00$. Then, we find

$$T_{min} = \left[\frac{\cancel{A}(1.00\times10^3\ \text{W/m}^2)}{\sigma \cancel{A}(1.00)}\right]^{1/4} = \left[\frac{1.00\times10^3\ \text{W/m}^2}{5.67\times10^{-8}\ \text{W/m}^2\cdot\text{K}^4}\right]^{1/4}$$

or $T_{min} = 364\ \text{K} = 91°\text{C}$ $(\approx 196°\text{F})$ ◊

53.   A 200-g block of copper at a temperature of 90°C is dropped into 400 g of water at 27°C. The water is contained in a 300-g glass container. What is the final temperature of the mixture?

Solution

Assuming that the mixture is thermally isolated from the environment, the total energy absorbed by the originally cooler parts of the mixture (water and glass) must equal the energy given up by the originally hotter material (copper). Hence, $Q_{\text{water}} + Q_{\text{glass}} = -Q_{\text{Cu}}$, or

$$\left(m_w c_w + m_g c_g \right)\left(\Delta T_w \right) = m_{\text{Cu}} c_{\text{Cu}} \left[-\left(\Delta T_{\text{Cu}} \right) \right]$$

$$\left[\left(0.400 \text{ kg} \right)\left(4\,186 \text{ J/kg} \cdot {}^\circ\text{C} \right) + \left(0.300 \text{ kg} \right)\left(837 \text{ J/kg} \cdot {}^\circ\text{C} \right) \right]\left(T_f - 27{}^\circ\text{C} \right) =$$

$$\left(0.200 \text{ kg} \right)\left(387 \text{ J/kg} \cdot {}^\circ\text{C} \right)\left[-\left(T_f - 90{}^\circ\text{C} \right) \right]$$

which yields a final temperature of

$$T_f = \frac{\left(77.4 \text{ J/}{}^\circ\text{C} \right)\left(90{}^\circ\text{C} \right) + \left(1.93 \times 10^3 \text{ J/}{}^\circ\text{C} \right)\left(27{}^\circ\text{C} \right)}{1.93 \times 10^3 \text{ J/}{}^\circ\text{C} + 77.4 \text{ J/}{}^\circ\text{C}} = 29{}^\circ\text{C}$$

◊

57. Water is being boiled in an open kettle that has a 0.500-cm-thick circular aluminum bottom with a radius of 12.0 cm. If the water boils away at a rate of 0.500 kg/min, what is the temperature of the lower surface of the bottom of the kettle? Assume that the top surface of the bottom of the kettle is at 100°C.

Solution

The energy that must be added to evaporate a mass m of water is $Q = mL_v$. If water evaporates at a rate $\Delta m/\Delta t$, the required rate of energy input is

$$\mathcal{P} = \frac{\Delta Q}{\Delta t} = \left(\frac{\Delta m}{\Delta t}\right)L_v = \left[\left(0.500 \ \frac{\text{kg}}{\text{min}}\right)\left(\frac{1 \ \text{min}}{60.0 \ \text{s}}\right)\right](2.26\times10^6 \ \text{J/kg}) = 1.88\times10^4 \ \text{J/s}$$

Assuming that a steady state exists inside the kettle, this must also equal the rate at which energy is transferred through the aluminum bottom of the kettle by conduction. Thus,

$$\mathcal{P} = kA\left(\frac{T_h - T_c}{L}\right) = 1.88\times10^4 \ \text{W}$$

The area of the kettle bottom is $A = \pi r^2 = \pi(0.120 \ \text{m})^2$, the thickness of the bottom is $L = 0.500 \ \text{cm} = 5.00\times10^{-3} \ \text{m}$, the thermal conductivity of aluminum is $k = 238 \ \text{J/s·m·°C}$, and we know that the temperature of the upper surface of the kettle bottom is $T_c = 100°\text{C}$. Thus, the temperature of the lower surface of the kettle bottom must be

$$T_h = T_c + \frac{(1.88\times10^4 \ \text{W})L}{kA} = 100°\text{C} + \frac{(1.88\times10^4 \ \text{W})(5.00\times10^{-3} \ \text{m})}{(238 \ \text{J/s·m·°C})\pi(0.120 \ \text{m})^2} = 109°\text{C} \qquad \Diamond$$

63.   A **flow calorimeter** is an apparatus used to measure the specific heat of a liquid. The technique is to measure the temperature difference between the input and output points of a flowing stream of the liquid while adding energy at a known rate.

(a) Start with the equations $Q = mc(\Delta T)$ and $m = \rho V$, and show that the rate at which energy is added to the liquid is given by the expression

$$\frac{\Delta Q}{\Delta t} = \rho c(\Delta T)\left(\frac{\Delta V}{\Delta t}\right)$$

(b) In a particular experiment, a liquid of density 0.72 g/cm^3 flows through the calorimeter at the rate of $3.5 \text{ cm}^3/\text{s}$. At steady state, a temperature difference of 5.8 °C is established between the input and output points when energy is supplied at the rate of 40 J/s. What is the specific heat of the liquid?

Solution

(a)   The energy ΔQ added to the volume ΔV of liquid that flows through the calorimeter in time Δt is

$$\Delta Q = (\Delta m)c(\Delta T) = \left[\rho(\Delta V)\right]c(\Delta T)$$

Thus, the rate of adding energy to the liquid is

$$\frac{\Delta Q}{\Delta t} = \rho c(\Delta T)\left(\frac{\Delta V}{\Delta t}\right)$$ ◊

where $(\Delta V/\Delta t)$ is the flow rate through the calorimeter.

(b)   From the result of part (a), the specific heat of the flowing liquid is

$$c = \frac{\Delta Q/\Delta t}{\rho(\Delta T)(\Delta V/\Delta t)} = \frac{40 \text{ J/s}}{(0.72 \text{ g/cm}^3)(5.8 \text{ °C})(3.5 \text{ cm}^3/\text{s})}$$

or $c = (2.7 \text{ J/g·°C})\left(\frac{10^3 \text{ g}}{1 \text{ kg}}\right) = 2.7 \times 10^3 \text{ J/kg·°C}$ ◊

Chapter 12

The Laws of Thermodynamics

NOTES FROM SELECTED CHAPTER SECTIONS

12.1 Work in Thermodynamics

The work done on a gas in a process that takes it from some initial state to some final state is the negative of the area under the curve on a PV diagram. (See the figure to the right).

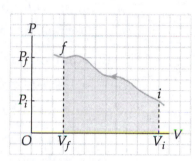

If the gas is compressed ($V_f < V_i$) the work done on the gas is positive.

If the gas expands ($V_f > V_i$) the work done on the gas is negative.

$$\begin{pmatrix} \text{Work} \\ \text{on gas} \end{pmatrix} = - \begin{pmatrix} \text{Area under} \\ \text{the curve} \end{pmatrix}$$

If the gas expands at constant pressure, called an **isobaric process**, then $W = -P(V_f - V_i)$.

The work done on a system depends on the process (path) by which the system goes from the initial to the final state; the quantity of work depends on the initial, final, and intermediate states of the system.

12.2 The First Law of Thermodynamics

In the **first law of thermodynamics**, $\Delta U = Q + W$, Q is the energy transferred to the system by heat and W is the work done on the system. Note that by convention, Q is positive when energy enters the system and negative when energy is removed from the system. Likewise, W can be positive or negative as mentioned earlier. The initial and final states must be equilibrium states; however, the intermediate states are, in general, non-equilibrium states since the thermodynamic coordinates undergo finite changes during the thermodynamic process.

12.3 Heat Engines and the Second Law of Thermodynamics

A **heat engine** is a device that converts internal energy to other useful forms, such as electrical and mechanical energy. A heat engine carries some working substance through a cyclic process during which (1) energy is transferred from a reservoir at a high temperature, (2) work is done by the engine, and (3) energy is expelled by the engine to a reservoir at a lower temperature.

The engine absorbs a quantity of energy, Q_h, from a hot reservoir, does work W_{eng} and then gives up energy Q_c to a cold reservoir. Because the working substance goes through a cycle, its initial and final internal energies are equal, so $\Delta U = 0$. Hence, from the first law of thermodynamics, *the work, done by a heat engine equals the net energy absorbed from the reservoirs:* $W_{eng} = Q_h - Q_c$.

If the working substance is a gas, the work done by the engine for a cyclic process is the area enclosed by the curve representing the process on a PV diagram.

The **thermal efficiency**, e, of a heat engine is the ratio of the work done by the engine to the energy absorbed at the higher temperature during one cycle.

The second law of thermodynamics can be stated as follows:

> It is impossible to construct a heat engine that, operating in a cycle, produces no other effect than the absorption of heat from a reservoir and the performance of an equal amount of work.

A process is **irreversible** if the system and its surroundings cannot be returned to their initial states. A process is **reversible** if the system passes from the initial to the final state through a succession of equilibrium states and can be returned to its initial condition along the same path.

The **Carnot cycle** is the most efficient cyclic process (or engine) operating between two given energy reservoirs with temperatures T_h and T_c. The *PV* diagram for the Carnot cycle, illustrated in the figure, includes **four reversible processes: two isothermal and two adiabatic.**

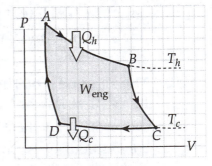

The Carnot Cycle

1. The process $A \rightarrow B$ is an isotherm (constant T), during which time the gas expands at constant temperature T_h and absorbs energy Q_h from the hot reservoir.

2. The process $B \rightarrow C$ is an adiabatic expansion ($Q = 0$), during which time the gas expands and cools to a temperature T_c.

3. The process $C \rightarrow D$ is a second isotherm, during which time the gas is compressed at constant temperature T_c, and expels energy Q_c to the cold reservoir.

4. The final process $D \rightarrow A$ is an adiabatic compression in which the gas temperature increases to a final temperature of T_h.

In practice, no working engine is 100% efficient, even when losses such as friction are neglected. One can obtain some theoretical limits on the efficiency of a real engine by comparison with the ideal Carnot engine. A **reversible engine** is one which will operate with the same efficiency in the forward and reverse directions. The Carnot engine is one example of a reversible engine.

All Carnot engines operating reversibly between T_h and T_c have the same efficiency given by Equation 12.16.

No real (irreversible) engine can have an efficiency greater than that of a reversible engine operating between the same two temperatures.

12.4   Entropy

Entropy is a quantity used to measure the **degree of disorder** in a system. For example, the molecules of a gas in a container at a high temperature are in a more disordered state (higher entropy) than the same molecules at a lower temperature.

When energy is transferred to a system by heat, the entropy increases. When energy is transferred out of a system by heat, the entropy decreases. In describing a thermodynamic process, the change in entropy is the important quantity; therefore the concept of entropy is most useful when a system undergoes a change in its state.

The second law of thermodynamics can be stated in terms of entropy as follows:

The total entropy of an isolated system always increases in time if the system undergoes an irreversible process. If an isolated system undergoes a reversible process, the total entropy remains constant.

EQUATIONS AND CONCEPTS

Equation 12.1 can be used to calculate the **work done on** a gas sample, if the pressure of the gas remains constant during a compression or an expansion. When ΔV is negative (compression), the work done on the gas is positive; when ΔV is positive (expansion), the work done on the gas is negative. The work done is equal to the negative of the area under the pressure-volume curve.

$$W = -P\Delta V \tag{12.1}$$

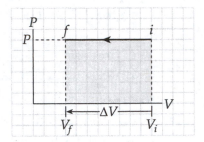

The **first law of thermodynamics** is stated in mathematical form in Equation 12.2. This law is a special case of the general principle of conservation of energy and includes changes in the internal energy of a system. The change in the internal energy of a system is equal to the sum of the energy transferred across the boundary by heat and the energy transferred by work. Q is positive when energy is added to the system by heat. W is positive when work is done on the system by its surroundings. *The values of both Q and W depend on the path or sequence of processes by which a system changes from an initial to a final state. The internal energy U depends only on the initial and final states of the system.*

$$\Delta U = Q + W \tag{12.2}$$

Q = energy added to system by heat

W = work done on system

ΔU = change in the internal energy of system

The following definitions will be important in describing some applications of the laws of thermodynamics.

In an isolated system $Q = W = 0$, so that $\Delta U = 0$. *The internal energy of an isolated system remains constant.*

Isolated system

A system which does not interact with its surroundings.

In a cyclic process, $\Delta U = 0$ and $Q = -W$. *The net work done on the gas per cycle equals the negative of the area enclosed by the path representing the process on a PV diagram.*

Cyclic process

A process for which the initial and final states are the same.

Equations 12.8 through 12.10 are valid only when the working substance approximates an ideal gas.

In an adiabatic process, $Q = 0$ and, therefore, the change in internal energy equals the work done on the system ($\Delta U = W$). *A system can undergo an adiabatic process if it is thermally insulated from its surroundings.*

Adiabatic process

A process in which no energy is transferred by heat.

$$PV^\gamma = \text{constant} \tag{12.8a}$$

$$\gamma = \frac{C_p}{C_v} \tag{12.8b}$$

In an isobaric process, the work done and the energy transferred by heat are both nonzero.

Isobaric process

A process which occurs at constant pressure.

$$Q = nC_p\Delta T \tag{12.6}$$

$$C_p = \frac{5}{2}R$$

At constant volume, $\Delta V = 0$, $W = 0$, and $\Delta U = Q$. *The net energy added by heat at constant volume goes into increasing the internal energy.*

Isovolumetric process

A process which occurs at constant volume.

$$Q = nC_v\Delta T \tag{12.9}$$

When a monoatomic ideal gas undergoes an isothermal process, $\Delta U = 0$ and $W = -Q$. *The work done on the system is equal to the negative of the energy added by heat.*

Isothermal process

A process which occurs at constant temperature.

$$W = -Q$$

$$W_{eng} = nRT \ln\left(\frac{V_f}{V_i}\right) \tag{12.10}$$

During each cycle:

(1) energy is transferred from a reservoir at a high temperature,

(2) work is done by the engine, and

(3) energy is expelled by the engine to a reservoir at a low temperature.

The net work, W_{eng}, done by a heat engine equals the net energy absorbed by the engine.

The **thermal efficiency** of a heat engine is the ratio of the work done to the energy absorbed at the higher temperature during one cycle of the process.

It is impossible to construct a heat engine that, operating in a cycle, produces no other effect than the absorption of energy from a reservoir and the performance of an equal amount of work.

The **Carnot cycle** is the most efficient cyclic process and the thermal efficiency of an ideal Carnot engine depends on the temperatures of the hot and cold reservoirs. All real engines are less efficient than the Carnot engine.

Entropy, S, is a thermodynamic variable which characterizes the degree of disorder in a system. *All physical processes tend toward a state of increasing entropy.*

The **change in entropy**, ΔS, as a system goes from an initial to a final state, is the ratio of the energy transferred to the system along a reversible path to the absolute temperature of the system.

Heat engine

A device that converts thermal energy into other forms of useful energy by carrying a substance through a cycle.

$$W_{eng} = |Q_h| - |Q_c| \qquad (12.11)$$

Q_h is the quantity of energy absorbed from the high temperature reservoir.

Q_c is the quantity of energy expelled to the low temperature reservoir.

$$e = \frac{W_{eng}}{|Q_h|} = \frac{|Q_h| - |Q_c|}{|Q_h|} = 1 - \frac{|Q_c|}{|Q_h|} \qquad (12.12)$$

Kelvin-Planck statement of the second law of thermodynamics

$$e_c = \frac{T_h - T_c}{T_h} = 1 - \frac{T_c}{T_h} \qquad (12.16)$$

$$\Delta S = \frac{\Delta Q_r}{T} \qquad (12.9)$$

The entropy of the universe increases in all natural processes. This is an alternate way of expressing the second law of thermodynamics.

REVIEW CHECKLIST

▷ Understand how work is defined when a system undergoes a change in state, and the fact that work depends on the path taken by the system. You should also know how to sketch processes on a *PV* diagram, and calculate work using these diagrams.

▷ State the first law of thermodynamics $(\Delta U = Q + W)$, and explain the meaning of the three forms of energy contained in this statement. Discuss the implications of the first law of thermodynamics as applied to (i) an isolated system, (ii) a cyclic process, (iii) an adiabatic process, and (iv) an isothermal process.

▷ Describe the processes via which an ideal heat engine goes through a **Carnot cycle.** Express the efficiency of an ideal heat engine (Carnot engine) as a function of work and energy exchange with its environment. Express the maximum efficiency of an ideal heat engine as a function of its input and output temperatures.

▷ Understand the concept of entropy. Define change in entropy for a system in terms of its energy gain or loss by heat, and its temperature. State the second law of thermodynamics as it applies to entropy changes in a thermodynamic system.

SOLUTIONS TO SELECTED END-OF-CHAPTER PROBLEMS

1. The only form of energy possessed by molecules of a monatomic ideal gas is translational kinetic energy. Using the results from the discussion of kinetic theory in Section 10.5, show that the internal energy of a monatomic ideal gas at pressure P and occupying volume V may be written as $U = \dfrac{3}{2}PV$.

Solution

The average kinetic energy per molecule in a monatomic ideal gas is $KE_{molecule} = \dfrac{3}{2}k_B T$, where T is the absolute temperature of the gas and the Boltzmann constant is

$$k_B = \frac{\text{universal gas constant}}{\text{Avogadro's number}} = \frac{R}{N_A}$$

If the gas contains N molecules, the total kinetic energy associated with random thermal motions is

$$KE = N(KE_{molecule}) = \frac{3}{2}N\left(\frac{R}{N_A}\right)T = \frac{3}{2}\left(\frac{N}{N_A}\right)RT$$

The total number of molecules, N, divided by the number of molecules in a mole, N_A, gives the number of moles of gas present, n.

Thus, the total kinetic energy becomes $\qquad\qquad KE = \dfrac{3}{2}nRT$

In an ideal gas, there are no intermolecular forces, so there are no potential energies contributing to the internal energy of the gas. The internal energy is therefore the same as the total kinetic energy,

or $\qquad\qquad\qquad\qquad\qquad\qquad\qquad\qquad U = \dfrac{3}{2}nRT$

Making use of the ideal gas law $\qquad\qquad\qquad\qquad PV = nRT$

this internal energy may be written as $\qquad\qquad\qquad U = \dfrac{3}{2}PV$ $\qquad\qquad\lozenge$

7. Gas in a container is at a pressure of 1.5 atm and a volume of 4.0 m³. What is the work done *on* the gas (a) if it expands at constant pressure to twice its initial volume? (b) if it is compressed at constant pressure to one-quarter its initial volume?

Solution

(a) In an isobaric process, the work done on the system is

$$W = -P(\Delta V) = -P\left(V_f - V_i\right)$$

Thus, if the constant pressure is $P = 1.5$ atm and $V_i = 4.0$ m³, the work done on the gas as it expands to twice its initial volume is

$$W = -(1.5 \text{ atm})\left(2V_i - V_i\right) = -1.5\left(1.013 \times 10^5 \text{ Pa}\right)\left(4.0 \text{ m}^3\right) = -6.1 \times 10^5 \text{ J} \qquad \Diamond$$

(b) If the gas is compressed to one-quarter of its initial volume, the work done on the gas is

$$W = -P(\Delta V) = -P\left(V_f - V_i\right) = -P\left(\frac{V_i}{4} - V_i\right) = +\frac{3}{4}PV_i$$

$$\text{or } W = +\frac{3}{4}\left[1.5\left(1.013 \times 10^5 \text{ Pa}\right)\right]\left(4.0 \text{ m}^3\right) = +4.6 \times 10^5 \text{ J} \qquad \Diamond$$

12.   A quantity of a monatomic ideal gas undergoes a process in which both its pressure and volume are doubled as shown in Figure P12.12. What is the energy absorbed by heat into the gas during this process? (**Hint:** See Problem 1.)

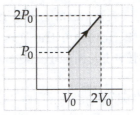

Solution

Figure P12.12

The work done on the gas is equal to the negative of the shaded area under the process curve on the *PV* diagram. This is a triangle on top of a rectangle, so

$$W = -\frac{1}{2}(2P_0 - P_0)(2V_0 - V_0) - P_0(2V_0 - V_0) = -1.5P_0V_0$$

Using Problem 12.1, the change in the internal energy of the gas is seen to be

$$\Delta U = U_f - U_0 = \frac{3}{2}P_fV_f - \frac{3}{2}P_0V_0 = \frac{3}{2}\left[(2P_0)(2V_0) - P_0V_0\right] = 4.5P_0V_0$$

The first law of thermodynamics, $\Delta U = Q + W$, then gives the energy absorbed by heat into the gas as

$$Q = \Delta U - W = 4.5P_0V_0 - (-1.5P_0V_0) = 6P_0V_0 \qquad \Diamond$$

15.   A gas expands from I to F in Figure P12.5. The energy added to the gas by heat is 418 J when the gas goes from I to F along the diagonal path. (a) What is the change in internal energy of the gas? (b) How much energy must be added to the gas by heat for the indirect path IAF to give the same change in internal energy?

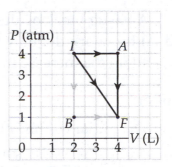

Solution

Figure P12.5

(a)   The area under the diagonal path IF is a triangle, of height 3.00 atm, sitting atop a rectangle with a height of 1.00 atm. The base of the triangle is the same as the base of the rectangle, 2.00 L. The work done on the gas during this process is then

$$W = -(\text{area under curve}) = -\frac{1}{2}(3.00 \text{ atm})(2.00 \text{ L}) - (1.00 \text{ atm})(2.00 \text{ L})$$

or $W = -5.00 \text{ atm} \cdot \text{L}\left(\frac{1.013 \times 10^5 \text{ N/m}^2}{1 \text{ atm}}\right)\left(\frac{10^{-3} \text{ m}^3}{1 \text{ L}}\right) = -506.5 \text{ J}$

Since the energy added to the gas by heat is $Q = 418 \text{ J}$ for this process, the first law of thermodynamics gives the change in the internal energy as

$$U_F - U_I = \Delta U = Q + W = 418 \text{ J} - 506.5 \text{ J} = -88.5 \text{ J} \qquad \lozenge$$

(b)   The area under the indirect path IAF is a simple rectangle, of height 4.00 atm and width 2.00 L. The work done on the gas during this process is

$$W = -(\text{area under curve}) = -(4.00 \text{ atm})(2.00 \text{ L}) = -8.00 \text{ atm} \cdot \text{L}$$

or $W = -8.00 \text{ atm} \cdot \text{L}\left(\frac{1.013 \times 10^5 \text{ N/m}^2}{1 \text{ atm}}\right)\left(\frac{10^{-3} \text{ m}^3}{1 \text{ L}}\right) = -810 \text{ J}$

The change in the internal energy of the gas between states F and I is the same as found above, $\Delta U = U_F - U_I = -88.5 \text{ J}$. The first law of thermodynamics, $\Delta U = Q + W$, then gives the energy added to the gas by heat in the process IAF as

$$Q = \Delta U - W = -88.5 \text{ J} - (-810 \text{ J}) = 722 \text{ J} \qquad \lozenge$$

21. A 5.0-kg block of aluminum is heated from 20°C to 90°C at atmospheric pressure. Find (a) the work done by the aluminum, (b) the amount of energy transferred to it by heat, and (c) the increase in its internal energy.

Solution

(a)   The initial volume of the aluminum block is

$$V_0 = \frac{m}{\rho} = \frac{5.0 \text{ kg}}{2.7 \times 10^3 \text{ kg/m}^3} = 1.85 \times 10^{-3} \text{ m}^3$$

As the block is heated, it undergoes an increase in volume given by

$$\Delta V = \beta V_0 \Delta T = (3\alpha) V_0 \Delta T$$
$$= 3 \left[24 \times 10^{-6} \ (°C)^{-1} \right] (1.85 \times 10^{-3} \text{ m}^3)(90°C - 20°C) = 9.3 \times 10^{-6} \text{ m}^3$$

and the work done **by** the aluminum is then

$$W_{\text{by block}} = +P(\Delta V) = (1.013 \times 10^5 \text{ Pa})(9.3 \times 10^{-6} \text{ m}^3) = 0.95 \text{ J} \qquad \Diamond$$

(b)   The energy transferred to the aluminum by heat during this process is

$$Q = mc_{\text{Al}}(\Delta T) = (5.0 \text{ kg})(900 \text{ J/kg} \cdot °C)(90°C - 20°C) = 3.2 \times 10^5 \text{ J} \qquad \Diamond$$

(c)   From the first law of thermodynamics, the change in the internal energy of the block is given by $\Delta U = Q + W$ where W is the work done **on** the aluminum block. Hence,

$$W = -W_{\text{by block}} = -0.95 \text{ J}$$

and $\Delta U = Q + W = 3.2 \times 10^5 \text{ J} - 0.95 \text{ J} = 3.2 \times 10^5 \text{ J}$ $\qquad \Diamond$

27.   One of the most efficient engines ever built is a coal-fired steam turbine engine in the Ohio River valley, driving an electric generator as it operates between 1 870 °C and 430 °C. (a) What is its maximum theoretical efficiency? (b) Its actual efficiency is 42.0%. How much mechanical power does the engine deliver if it absorbs 1.40×10^5 J of energy each second from the hot reservoir?

Solution

(a)   The maximum theoretical efficiency is that of a Carnot engine operating between the specified reservoirs. The **absolute** temperatures of the given hot and cold reservoirs are:

$$T_h = 1\,870 + 273 = 2\,143 \text{ K} \qquad \text{and} \qquad T_c = 430 + 273 = 703 \text{ K}$$

The Carnot efficiency is then

$$e_c = \frac{T_h - T_c}{T_h} = 1 - \frac{T_c}{T_h} = 1 - \frac{703 \text{ K}}{2\,143 \text{ K}} = 0.672 \quad (\text{or } 67.2\%) \qquad \lozenge$$

(b)   The actual efficiency of a heat engine is defined as $e = W_{eng}/|Q_h|$, where W_{eng} is the work done by the engine during some interval and $|Q_h|$ is the energy absorbed from the hot reservoir during the same interval.

This engine absorbs $|Q_h| = 1.40 \times 10^5$ J from the hot reservoir each second and has an actual efficiency of $e = 0.420$ (42.0%). Therefore the work done by the engine each second (or the power delivered) is

$$\mathcal{P} = \frac{W_{eng}}{\Delta t} = \frac{e|Q_h|}{\Delta t} = \frac{(0.420)(1.40 \times 10^5 \text{ J})}{1.00 \text{ s}} = 5.88 \times 10^4 \text{ W} = 58.8 \text{ kW} \qquad \lozenge$$

32. A heat engine operates in a Carnot cycle between 80.0°C and 350°C. It absorbs 21 000 J of energy per cycle from the hot reservoir. The duration of each cycle is 1.00 s. (a) What is the mechanical power output of this engine? (b) How much energy does it expel in each cycle by heat?

Solution

(a) The absolute temperatures of the reservoirs used by this engine are

$$T_c = 80.0 + 273 = 353 \text{ K} \quad \text{and} \quad T_h = 350 + 273 = 623 \text{ K}$$

and the efficiency when operating on the Carnot cycle is

$$e_c = 1 - \frac{T_c}{T_h} = 1 - \frac{353 \text{ K}}{623 \text{ K}} = 0.433 \quad \text{(or 43.3\%)}$$

If the engine absorbs energy at the rate of $|Q_h| = 21\,000 \text{ J} = 21.0 \text{ kJ}$ each cycle, and the duration of a cycle is one second, the power output of the engine is

$$\mathcal{P} = \frac{W_{eng}}{\Delta t} = \frac{e_c |Q_h|}{\Delta t} = \frac{(0.433)(21.0 \text{ kJ})}{1.00 \text{ s}} = 9.10 \text{ kW} \qquad \Diamond$$

(b) The energy that is expelled by heat during each cycle is

$$|Q_c| = |Q_h| - W_{eng} = |Q_h| - e_c |Q_h| = |Q_h|(1 - e_c)$$

or $\quad |Q_c| = (21.0 \text{ kJ})(1 - 0.433) = 11.9 \text{ kJ}$ $\qquad \Diamond$

35. A freezer is used to freeze 1.0 L of water completely into ice. The water and the freezer remain at a constant temperature of $T = 0°C$. Determine (a) the change in the entropy of the water and (b) the change in the entropy of the freezer.

Solution

The change in the entropy of a thermodynamic system during a reversible, constant temperature, process is given by

$$\Delta S = \frac{\Delta Q_r}{T}$$

where ΔQ_r is the energy added to the system by heat and T is the **absolute** temperature of the system during this process.

As 1.0-L of liquid water at 0 °C is converted to ice, the quantity of energy transferred by heat **from** the water and **to** the freezer is

$$|\Delta Q_r| = mL_f = (\rho V)L_f = \left[\left(10^3 \text{ kg/m}^3\right)(1.0 \text{ L})\left(\frac{10^{-3} \text{ m}^3}{1 \text{ L}}\right) \right](3.33 \times 10^5 \text{ J/kg})$$

or $\qquad |\Delta Q_r| = 3.3 \times 10^5 \text{ J} = 330 \text{ kJ}$

The constant temperature of both the water and freezer during this process is $T = 0 \text{ °C} = 273 \text{ K}$.

(a) Energy is transferred from the water during the freezing process, so $\left(\Delta Q_r\right)_{\text{water}} < 0$ and the change in the entropy of the water is

$$\Delta S_{\text{water}} = \frac{\left(\Delta Q_r\right)_{\text{water}}}{T} = \frac{-330 \text{ kJ}}{273 \text{ K}} = -1.2 \text{ kJ/K} \qquad \diamond$$

(b) Energy is transferred to the freezer by heat, so $\left(\Delta Q_r\right)_{\text{freezer}} > 0$

and $\qquad \Delta S_{\text{freezer}} = \frac{\left(\Delta Q_r\right)_{\text{freezer}}}{T} = \frac{+330 \text{ kJ}}{273 \text{ K}} = +1.2 \text{ kJ/K} \qquad \diamond$

Note that in this **reversible** process, the total entropy of the isolated system consisting of (freezer plus water) is constant,

or $\qquad \Delta S_{\text{isolated system}} = \Delta S_{\text{water}} + \Delta S_{\text{freezer}} = 0$

39. The surface of the Sun is approximately at 5 700 K, and the temperature of Earth's surface is approximately 290 K. What entropy change occurs when 1 000 J of energy is transferred by heat from the Sun to Earth?

Solution

A quantity of energy $|Q|$ is transferred **from** the Sun at absolute temperature T_{Sun}. The change in the entropy of the Sun during this process is

$$\Delta S_{Sun} = \frac{Q_r}{T_{Sun}} = \frac{-|Q|}{T_{Sun}}$$

This same quantity of energy is transferred **to** Earth at absolute temperature T_{Earth}, and the change in the entropy of Earth for this process is

$$\Delta S_{Earth} = \frac{Q_r}{T_{Earth}} = \frac{+|Q|}{T_{Earth}}$$

The total change in entropy of the Universe during this energy exchange is then

$$\Delta S_{total} = \Delta S_{Sun} + \Delta S_{Earth} = \frac{-|Q|}{T_{Sun}} + \frac{+|Q|}{T_{Earth}} = |Q|\left(\frac{1}{T_{Earth}} - \frac{1}{T_{Sun}}\right)$$

Thus, if $|Q| = 1\,000$ J, $T_{Sun} = 5\,700$ K, and $T_{Earth} = 290$ K, the total change in entropy is

$$\Delta S_{total} = (1\,000 \text{ J})\left(\frac{1}{290 \text{ K}} - \frac{1}{5\,700 \text{ K}}\right) = +3.27 \text{ J/K} \qquad \Diamond$$

Note that we have treated both the withdrawal of energy from the Sun and the addition of energy to Earth as constant temperature processes. The justification for this is that the energy transferred $(|Q| = 1\,000 \text{ J})$ is very small in comparison to the total internal energy of each reservoir (Sun and Earth) involved in the transfer.

47.   Find the change in temperature of a river due to the exhausted energy from a nuclear power plant. Assume that the input power to the boiler in the plant is 25×10^8 W, the efficiency of use of this power is 30%, and the river flow rate is 9.0×10^6 kg/min.

Solution

We treat the power plant as a heat engine, taking in energy from a high temperature reservoir at a rate of $|Q_h|/\Delta t = 25 \times 10^8$ W, producing useful power at a rate $W_{eng}/\Delta t$ with an efficiency of $e = 0.30$, and exhausting energy to the low temperature reservoir (river) at a rate $|Q_c|/\Delta t$.

The rate at which waste energy is exhausted is

$$\frac{|Q_c|}{\Delta t} = \frac{|Q_h|}{\Delta t} - \frac{W_{eng}}{\Delta t} = \frac{|Q_h|}{\Delta t} - \frac{e|Q_h|}{\Delta t} = \frac{|Q_h|}{\Delta t}(1-e)$$

or

$$\frac{|Q_c|}{\Delta t} = (25 \times 10^8 \text{ W})(1-0.30) = \left(1.8 \times 10^9 \frac{\text{J}}{\text{s}}\right)\left(\frac{60 \text{ s}}{1 \text{ min}}\right) = 1.1 \times 10^{11} \text{ J/min}$$

If the quantity of energy exhausted each minute $|Q_c| = 1.1 \times 10^{11}$ J is input to the 9.0×10^6 kg of water that flows through the power plant each minute, the rise in temperature of that water will be

$$\Delta T = \frac{|Q_c|}{mc_{water}} = \frac{1.1 \times 10^{11} \text{ J}}{(9.0 \times 10^6 \text{ kg})(4\,186 \text{ J/kg} \cdot {}^\circ\text{C})} = 2.8{}^\circ\text{C} \qquad \Diamond$$

51. A substance undergoes the cyclic process shown in Figure P12.51. Work output occurs along path AB, while work input is required along path BC, and no work is involved in the constant volume process CA. Energy transfers by heat occur during each process involved in the cycle. (a) What is the work output during process AB? (b) How much work input is required during process BC? (c) What is the net energy input Q during this cycle?

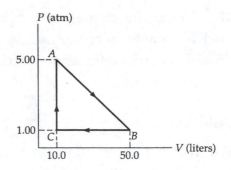

Figure P12.51

Solution

(a) The work done **on** the system (work input) during process AB equals the negative of the area under the process curve on the PV diagram. This consists of a triangular area sitting atop a rectangular area. Thus,

$$W_{AB} = -\left[(\text{area of triangle}) + (\text{area of rectangle})\right]$$

$$= -\frac{1}{2}\left[(5.00-1.00)\ \text{atm}\right]\left[(50.0-10.0)\ \text{L}\right] - (1.00\ \text{atm})\left[(50.0-10.0)\ \text{L}\right]$$

or $W_{AB} = -(120\ \text{atm} \cdot \text{L})\left(\dfrac{1.013\times10^{5}\ \text{Pa}}{1\ \text{atm}}\right)\left(\dfrac{10^{-3}\ \text{m}^{3}}{1\ \text{L}}\right) = -1.22\times10^{4}\ \text{J} = -12.2\ \text{kJ}$

The work done **by** the system (work output) during process AB is the negative of the work input computed above. That is,

$$W_{output} = -W_{AB} = -(-12.2\ \text{kJ}) = +12.2\ \text{kJ}$$ ◊

(b) The work done **on** the system (work input) during the process BC is the negative of the rectangular area under this process curve on the PV diagram. That is,

$$W_{BC} = -(\text{area under curve}) = -P(V_C - V_B) = -(1.00\ \text{atm})\left[(10.0-50.0)\ \text{L}\right]$$

or $W_{BC} = (+40.0\ \text{atm} \cdot \text{L})\left(\dfrac{1.013\times10^{5}\ \text{Pa}}{1\ \text{atm}}\right)\left(\dfrac{10^{-3}\ \text{m}^{3}}{1\ \text{L}}\right) = 4.05\times10^{3}\ \text{J} = 4.05\ \text{kJ}$ ◊

(c) The net change in the internal energy of a system is $\Delta U_{net} = 0$ for any cyclic process. Thus, for the entire cycle, the first law of thermodynamics gives

$$\Delta U_{net} = Q_{net} + W_{net} = 0 \quad \text{or} \quad Q_{net} = -W_{net} = -\left(W_{AB} + W_{BC} + W_{CA}\right)$$

and the net energy input as heat during the full cycle is

$$Q_{net} = -\left[-12.2\ \text{kJ} + 4.05\ \text{kJ} + 0\right] = +8.15\ \text{kJ}$$ ◊

57. Suppose a heat engine is connected to two energy reservoirs, one a pool of molten aluminum at 660 °C and the other a block of solid mercury at −38.9 °C. The engine runs by freezing 1.00 g of aluminum and melting 15.0 g of mercury during each cycle. The latent heat of fusion of aluminum is 3.97×10^5 J/kg, and that of mercury is 1.18×10^4 J/kg. (a) What is the efficiency of this engine? (b) How does the efficiency compare with that of a Carnot engine?

Solution

(a) During each cycle, the energy transferred to the engine from the molten aluminum is

$$|Q_h| = m_{Al}\left(L_f\right)_{Al} = \left(1.00 \times 10^{-3}\ \text{kg}\right)\left(3.97 \times 10^5\ \text{J/kg}\right) = 397\ \text{J}$$

Also, during each cycle, the energy exhausted from the engine to the frozen mercury is

$$|Q_c| = m_{Hg}\left(L_f\right)_{Hg} = \left(15.0 \times 10^{-3}\ \text{kg}\right)\left(1.18 \times 10^4\ \text{J/kg}\right) = 177\ \text{J}$$

Thus, the efficiency of the heat engine is given by

$$e = \frac{W_{eng}}{|Q_h|} = \frac{|Q_h| - |Q_c|}{|Q_h|} = 1 - \frac{|Q_c|}{|Q_h|} = 1 - \frac{177\ \text{J}}{397\ \text{J}} = 0.554 \quad (\text{or } 55.4\%) \qquad \Diamond$$

(b) The efficiency of a Carnot engine operating between reservoirs having temperatures of $T_h = 660\ °C = 933$ K and $T_c = -38.9\ °C = 234$ K is

$$e_c = \frac{T_h - T_c}{T_h} = 1 - \frac{T_c}{T_h} = 1 - \frac{234\ \text{K}}{933\ \text{K}} = 0.749 \quad (\text{or } 74.9\%) \qquad \Diamond$$

Thus, the actual efficiency of this engine is $\left(\dfrac{0.554}{0.749}\right) \times 100\% = 74.0\%$ of the theoretical maximum possible efficiency.

NOTES FROM SELECTED CHAPTER SECTIONS

13.1 Hooke's Law

13.2 Elastic Potential Energy

Simple harmonic motion occurs when the net force along the direction of motion is a Hooke's law type of force; that is, when the net force is proportional to the displacement and in the opposite direction.

It is necessary to define a few terms relative to harmonic motion:

- **The amplitude,** A, is the maximum distance that an object moves away from its equilibrium position. In the absence of friction, an object will continue in simple harmonic motion and reach a maximum displacement equal to the amplitude on each side of the equilibrium position during each cycle.

- **The period, T,** is the time it takes the object to execute one complete cycle of the motion.

- **The frequency,** f, is the number of cycles or vibrations per unit of time.

Oscillatory motions are exhibited by many physical systems such as a mass attached to a spring, a pendulum, atoms in a solid, stringed musical instruments, and electrical circuits driven by a source of alternating current. *Simple harmonic motion of a mechanical system corresponds to the oscillation of an object between two points for an indefinite period of time, with no loss in mechanical energy.*

An object exhibits simple harmonic motion if the net external force acting on it is a linear restoring force.

13.4   Position, Velocity, and Acceleration as Functions of Time

The position (x), velocity (v), and acceleration (a) of an object moving with simple harmonic motion are shown in the three graphs below.

In this particular case, the object was released from rest when it was a maximum distance (amplitude) from the equilibrium position.

Shown here, from top to bottom, are graphs of displacement, velocity, and acceleration versus time for an object moving with simple harmonic motion under the initial conditions that $x_0 = A$ and $v_0 = 0$ at $t = 0$.

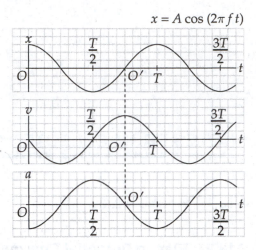

$x = A \cos(2\pi f t)$

The most common system which undergoes simple harmonic motion is the **mass-spring system** shown in the figure at the right. The mass is assumed to move on a horizontal, frictionless surface. The point $x = 0$ is the equilibrium position of the mass; that is, the point where the mass would reside if left undisturbed. In this position, there is no horizontal force on the mass. When the mass is displaced a distance x from its equilibrium position, the spring produces a linear restoring force given by Hooke's law, $F = -kx$. The parameter k is the force constant of the spring and has SI units of N/m. The minus sign means that $\vec{F}$ is to the left when the displacement x is positive, whereas $\vec{F}$ is to the right when x is negative. *The direction of the force $\vec{F}$ is always toward the equilibrium position.*

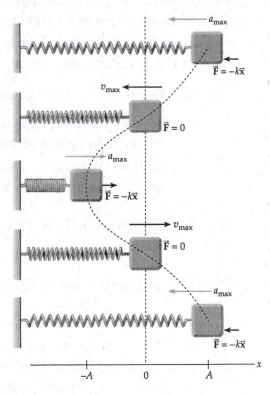

13.5 Motion of a Pendulum

A **simple pendulum** consists of a mass m attached to a light string of length L as shown in the figure. When the angular displacement θ is small during the entire motion (less than about 15°), the pendulum exhibits simple harmonic motion. In this case, the resultant force acting on the mass m equals the component of weight tangent to the circular path followed by the mass. The magnitude of the resultant force equals $mg\sin\theta$.

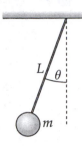

Since this force is always directed toward $\theta = 0$, it corresponds to a restoring force.

The period depends only on the length of the pendulum and the acceleration due to gravity. The period does not depend on mass, so we conclude that all simple pendula of equal length oscillate with the same frequency or period at the same location.

13.6 Damped Oscillations

Damped oscillations occur in realistic systems in which retarding forces such as friction are present. These forces will reduce the amplitudes of the oscillations with time, since mechanical energy is continually transferred from the oscillating system. Depending on the value of the frictional (retarding) force three distinct types of damping can be identified:

Underdamped: In this case, the retarding force is small compared to the restoring force. Oscillations continue with the frequency of the undamped system (passing through the equilibrium point twice during each oscillation). The amplitude of the motion decreases exponentially with time until the oscillations cease at zero amplitude.

Critically damped: With an increase in the retarding force the system does not oscillate and returns to the equilibrium position in the shortest possible time without passing through the equilibrium point.

Overdamped: This mode is similar to critical damping and occurs with further increase in the frictional force. In this case the system returns to equilibrium without passing through the equilibrium point but requires a longer time to do so.

It is possible to compensate for the energy lost in a damped oscillator by adding an additional driving force that does positive work on the system. This additional energy supplied to the system must equal the energy lost due to friction to maintain constant amplitude.

13.7 Wave Motion

The production of mechanical waves requires: (1) an **elastic medium** which can be disturbed, (2) **an energy source** to provide a disturbance or deformation in the medium, and (3) a **physical mechanism** by way of which adjacent portions of the medium can influence each other. The three parameters important in characterizing waves are (1) **wavelength**, (2) **frequency**, and (3) **wave velocity**.

Transverse waves are those in which particles of the disturbed medium move along a direction which is perpendicular to the direction of the wave velocity. For **longitudinal waves,** the particles of the medium undergo displacements which are parallel to the direction of wave motion.

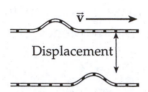

Transverse Wave

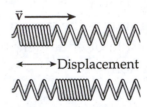

Longitudinal Wave

13.8 Frequency, Amplitude, and Wavelength

Consider a wave traveling in the x direction on a very long string. Each particle along the string oscillates in simple harmonic motion along the y direction with a **frequency** equal to

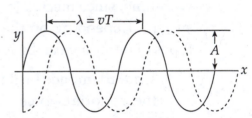

the frequency of the source producing the vibration. The maximum distance the string is displaced above or below the equilibrium value is called the **amplitude,** A, of the wave. The distance between two successive points along the string having maximum upper displacement is called the **wavelength,** λ. The number of crests or peaks of the wave passing a fixed point each second is the wave frequency (or harmonic frequency). The wave will advance along the string a distance of one wavelength in a time interval equal to one **period** of vibration, T; the time required for any point in the medium to complete one full cycle of vibration.

13.9 The Speed of Waves on Strings

For linear waves, the **speed of mechanical waves** depends only on the physical properties of the medium through which the disturbance travels. In the case of **waves on a string,** the velocity depends on the tension in the string and its mass per unit length (linear mass density).

13.10 Interference of Waves

If two or more waves are moving through a medium, the **resultant wave function** is the algebraic sum of the wave functions of the individual waves. Two traveling waves can pass through each other without being destroyed or altered.

13.11 Reflection of Waves

Whenever a traveling wave reaches a boundary, **reflection** occurs; part or all of the wave is reflected back through the original medium . If the wave is traveling along a string and is reflected from a "fixed" end, the reflected pulse is inverted. By contrast, a pulse is reflected without inversion at the "free" end of a string.

EQUATIONS AND CONCEPTS

The force exerted by a spring on a mass attached to the spring and displaced a distance x from the unstretched position is given by **Hooke's law**. The force constant, k, is always positive and has a value which corresponds to the relative stiffness of the spring. The negative sign means that the force exerted by the string on the mass is always directed opposite the displacement; the force is a restoring force, always directed toward the equilibrium position.

$$F_s = -kx \qquad (13.1)$$

In simple harmonic motion $\vec{F} = -k\vec{x}$

An object exhibits **simple harmonic motion** when the net force along the direction of motion is proportional to the displacement and oppositely directed.

This equation gives the acceleration of an object in simple harmonic motion as a function of position. Note that when the oscillating mass is at the equilibrium position ($x=0$), the acceleration $a=0$. The acceleration has its maximum magnitude when the displacement of the mass is maximum, $x=\pm A$ (amplitude).

$$a = -\left(\frac{k}{m}\right)x \qquad (13.2)$$

Acceleration as a function of position

Work must be done by an external applied force in order to stretch or compress a spring. This work results in energy, called **elastic potential energy**, being stored in the spring.

$$PE_s \equiv \tfrac{1}{2}k\,x^2 \qquad (13.3)$$

The spring force is conservative; hence, in the absence of friction or other non-conservative forces, the total mechanical energy (kinetic, gravitational potential, and elastic potential) of the spring-mass system remains constant.

$$(KE + PE_g + PE_s)_i = \\ (KE + PE_g + PE_s)_f \qquad (13.4)$$

The **speed of an object in simple harmonic motion** is a maximum at $x=0$; the speed is zero when the mass is at the points of maximum displacement ($x=\pm A$).

$$v = \pm\sqrt{\frac{k}{m}(A^2 - x^2)} \qquad (13.6)$$

Speed as a function of position

The **period** of an object in simple harmonic motion is the time required to complete a full cycle of its motion.

$$T = 2\pi\sqrt{\frac{m}{k}} \qquad (13.8)$$

The **harmonic frequency** f is the number of cycles or oscillations completed per unit time and is the reciprocal of the period. The units of frequency are hertz (Hz).

$$f = \frac{1}{T} \qquad (13.9)$$

$$f = \frac{1}{2\pi}\sqrt{\frac{k}{m}} \qquad (13.10)$$

The quantity ω is the **angular frequency** measured in radians per second.

$$\omega = 2\pi f \qquad (13.11)$$

Equations 13.14a,b,c represent the position, velocity, and acceleration of an object moving in simple harmonic motion along the x-axis as a function of time.

$$x = A\cos(\omega t) \tag{13.12}$$

$$x = A\cos(2\pi f t) \tag{13.14a}$$

$$v = -A\omega\sin(2\pi f t) \tag{13.14b}$$

$$a = -A\omega^2\cos(2\pi f t) \tag{13.14c}$$

In Equations 13.12 and 13.14a, $x = A$, when $t = 0$ (since $\cos 0 = 1$). Therefore, the particular form of the position equations shown above assumes that the vibrating object is at the point of maximum displacement when $t = 0$.

The **period of a simple pendulum** depends only on its length, L, and the acceleration due to gravity, g. The period does not depend on the mass; and to a good approximation, the period does not depend on the amplitude, θ_{max}, within the range of small amplitudes.

$$T = 2\pi\sqrt{\frac{L}{g}} \tag{13.15}$$

The **wave speed**, v, is the rate at which a disturbance or pulse moves along the direction of travel of the wave.

Wave speed, v

$$v = f\lambda \tag{13.17}$$

The **wave speed in a stretched string** depends on the tension in the string and its linear density (mass per unit length). *For any mechanical wave, the speed depends only on the properties of the medium through which the wave travels.*

$$v = \sqrt{\frac{F}{\mu}} \tag{13.18}$$

REVIEW CHECKLIST

▷ Describe the general characteristics of a system in simple harmonic motion; and define amplitude, period, frequency, and displacement.

▷ Define the following terms relating to wave motion: frequency, wavelength, velocity, and amplitude; and express a given harmonic wave function in several alternative forms involving different combinations of the wave parameters: wavelength, period, phase velocity, angular frequency, and harmonic frequency.

▷ Given a specific wave function for a harmonic wave, obtain values for the characteristic wave parameters: A, ω, and f.

▷ Make calculations which involve the relationships between wave speed and the inertial and elastic characteristics of a string through which the disturbance is propagating.

▷ Define and describe the following wave associated phenomena: superposition, phase, interference, and reflection.

SOLUTIONS TO SELECTED END-OF-CHAPTER PROBLEMS

3. A ball dropped from a height of 4.00 m makes a perfectly elastic collision with the ground. Assuming no mechanical energy is lost due to air resistance, (a) show that the motion is periodic and (b) determine the period of the motion. (c) Is the motion simple harmonic? Explain.

Solution

(a) Since the collision is perfectly elastic $(KE_{after} = KE_{before})$ and the recoil of the Earth may be neglected, the speed of the ball as it leaves the ground after a collision is the same as its speed immediately before the collision. The ball rebounds to the height from which it was initially dropped, comes to rest, and then falls again.

The ball repeats this motion continuously, and thus, undergoes a periodic motion.   ◊

(b) The period (time for one complete cycle) of this motion is the total elapsed time from when the ball starts at rest 4.00 m above ground until it returns to rest at 4.00 m high following a bounce. For the downward motion,

$$v_0 = 0 \qquad\qquad \Delta y = -4.00 \text{ m} \qquad\qquad a_y = -g = -9.80 \text{ m/s}^2$$

Thus, $\quad \Delta y = v_0 t + \frac{1}{2} a_y t^2 \quad$ gives the time for the ball to reach the ground:

$$t_1 = \sqrt{\frac{2(\Delta y)}{a_y}} = \sqrt{\frac{2(-4.00 \text{ m})}{-9.80 \text{ m/s}^2}} = 0.904 \text{ s}$$

The speed of the ball just before impact is

$$v_1 = v_0 + a_y t_1 = 0 + (-9.80 \text{ m/s}^2)(0.904 \text{ s}) = -8.85 \text{ m/s}$$

On the upward flight, the initial velocity is $v_2 = -v_1 = +8.85 \text{ m/s}$, $v_f = 0$, and $\Delta y = +4.00 \text{ m}$. Therefore, the average velocity for this part of the motion is $v_{av} = \frac{1}{2}(v_2 + v_f) = 4.43 \text{ m/s}$, and the time to return to the initial height is $t_2 = \Delta y / v_{av} = (4.00 \text{ m})/(4.43 \text{ m/s}) = 0.904 \text{ s}$. The total time for this complete cycle of the motion is then

$$T = t_1 + t_2 = 1.81 \text{ s} \qquad\qquad ◊$$

(c)   Simple harmonic motion is motion produced by a Hooke's law type force. Such a force is proportional to the displacement from an equilibrium position and is in the direction opposite to the displacement. This ball has a constant downward force, $w = mg$, acting on it while it is in the air. When the ball is at ground level, an upward impulsive force acts on it. These forces are not Hooke's law type forces. Thus, the motion **is not** simple harmonic.   ◊

---

7.   A slingshot consists of a light leather cup containing a stone. The cup is pulled back against two parallel rubber bands. It takes a force of 15 N to stretch either one of these bands 1.0 cm. (a) What is the potential energy stored in the two bands together when a 50-g stone is placed in the cup and pulled back 0.20 m from the equilibrium position? (b) With what speed does the stone leave the slingshot?

Solution

(a)   Assume the rubber bands obey Hooke's law. Then, the force constant of either one of the bands is given by

$$k = \frac{F}{|\Delta x|} = \frac{15 \text{ N}}{1.0 \times 10^{-2} \text{ m}} = 1.5 \times 10^{3} \text{ N/m}$$

Thus, the total elastic potential energy stored in the two bands when the cup is pulled back a distance $x_i = 0.20$ m is

$$PE_i = 2\left(PE_{\substack{one \\ band}} \right) = 2\left(\frac{1}{2} kx_i^2 \right) = \left(1.5 \times 10^3 \text{ } \frac{\text{N}}{\text{m}} \right)(0.20 \text{ m})^2 = 60 \text{ J}$$ ◊

(b)   Consider the system consisting of the stone and the two bands. With the initial state being just before the cup is released and the final state being the instant the stone leaves the cup, assumed to occur at the equilibrium position (bands relaxed), conservation of mechanical energy yields:

$$KE_f + PE_f = KE_i + PE_i \implies \frac{1}{2} m_{stone} v_{max}^2 + 0 = 0 + PE_i$$

or $$v_{max} = \sqrt{\frac{2(PE_i)}{m_{stone}}} = \sqrt{\frac{2(60 \text{ J})}{50 \times 10^{-3} \text{ kg}}} = 49 \text{ m/s}$$ ◊

---

17.   At an outdoor market, a bunch of bananas is set into oscillatory motion with an amplitude of 20.0 cm on a spring with a force constant of 16.0 N/m. It is observed that the maximum speed of the bunch of bananas is 40.0 cm/s. What is the weight of the bananas in newtons?

Solution

The bananas have maximum speed as they pass through the equilibrium position, $x = 0$, where the elastic potential energy stored in the spring is zero. When the bananas are at maximum displacement from the equilibrium position, $x = \pm A$, they are momentarily at rest $(v = 0)$. Since the Hooke's law type force exerted on the bananas by the spring is a conservative force, the total mechanical energy of the spring-bananas system is constant. Applying conservation of energy from when the bananas are at the equilibrium position to when they are at maximum displacement from equilibrium gives

$$KE_i + PE_i = KE_f + PE_f \Rightarrow \frac{1}{2}mv_{max}^2 + 0 = 0 + \frac{1}{2}kA^2$$

Thus, the mass of the bananas is

$$m = \frac{kA^2}{v_{max}^2} = \frac{(16.0 \text{ N/m})(0.200 \text{ m})^2}{(0.400 \text{ m/s})^2} = 4.00 \text{ kg}$$

and their weight is $F_g = mg = (4.00 \text{ kg})(9.80 \text{ m/s}^2) = 39.2 \text{ N}$ ◊

You should observe that in applying conservation of energy to this vertical spring-mass system, we have ignored the gravitational potential energy of the mass (bananas). It can be shown that it is valid to do this, and still have gravitational potential energy properly accounted for, if the zero point of the elastic potential energy is taken at the position where the mass will hang in equilibrium on the end of the spring rather than at the position where the spring is completely relaxed and unstretched.

23.   A spring stretches 3.9 cm when a 10-g object is hung from it. The object is replaced with a block of mass 25 g that oscillates in simple harmonic motion. Calculate the period of motion.

Solution

A simple harmonic oscillator moves under the influence of a Hooke's law force, $\vec{F}_s = -k\vec{x}$ where $\vec{x}$ is the displacement from equilibrium. The spring constant is given by $k = \left| \vec{F}_s \right| / \left| \vec{x} \right| = F_s / x$.

When a 10-g object hangs from the spring, the force stretching the spring is $F_s = mg = (10 \times 10^{-3} \text{ kg})(9.80 \text{ m/s}^2) = 0.098 \text{ N}$. If this produces an elongation of $x = 3.9 \text{ cm} = 0.039 \text{ m}$, the spring constant is

$$k = \frac{F_s}{x} = \frac{0.098 \text{ N}}{0.039 \text{ m}} = 2.5 \text{ N/m}$$

When a total mass of 25 g is attached to this spring and set into oscillation, the period of the resulting simple harmonic motion will be

$$T = 2\pi \sqrt{\frac{m}{k}} = 2\pi \sqrt{\frac{25 \times 10^{-3} \text{ kg}}{2.5 \text{ N/m}}} = 0.63 \text{ s}$$

◊

29. Given that $x = A\cos(\omega t)$ is a sinusoidal function of time, show that v (velocity) and a (acceleration) are also sinusoidal functions of time. (**Hint:** Use Equations 13.6 and 13.2.)

Solution

The total energy of a simple harmonic oscillator is

$$E = KE + PE_s = \tfrac{1}{2}mv^2 + \tfrac{1}{2}kx^2$$

Since the force acting on the oscillating mass is a **conservative** force ($\vec{F}_s = -k\vec{x}$), the total energy is constant. At the maximum displacement from equilibrium ($x = A$), the speed is $v = 0$, so the total energy is $E = 0 + \tfrac{1}{2}kA^2 = \tfrac{1}{2}kA^2$. Thus, at arbitrary displacement x, we have $\tfrac{1}{2}mv^2 + \tfrac{1}{2}kx^2 = \tfrac{1}{2}kA^2$ which gives the velocity as

$$v = \pm\sqrt{\frac{k}{m}\left(A^2 - x^2\right)} = \pm\sqrt{\omega^2\left(A^2 - x^2\right)} = \pm\omega\sqrt{A^2 - x^2} \qquad \textbf{[Equation 13.6]}$$

From Newton's second law, the acceleration is $\vec{a} = \vec{F}/m$. Thus, when $\vec{F} = \vec{F}_s = -k\vec{x}$, the acceleration is given by $a = -\dfrac{k}{m}x = -\omega^2 x$. $\qquad$ **[Equation 13.2]**

For a simple harmonic oscillator, the position is a sinusoidal function of time, $x = A\cos(\omega t)$. The velocity is then

$$v = \pm\omega\sqrt{A^2 - x^2} = \pm\omega A\sqrt{1 - \cos^2(\omega t)} = \pm\omega A\sin(\omega t)$$

which is another sinusoidal function of time. $\qquad\qquad\qquad\qquad\qquad\qquad\qquad\quad \lozenge$

The acceleration of the simple harmonic oscillator is given by

$$a = -\omega^2 x = -\left(\omega^2 A\right)\cos(\omega t), \text{ also a sinusoidal function of time.} \qquad\qquad \lozenge$$

35. The free-fall acceleration on Mars is 3.7 m/s^2. (a) What length of pendulum has a period of 1 s on Earth? What length of pendulum would have a 1-s period on Mars? (b) An object is suspended from a spring with force constant 10 N/m. Find the mass suspended from this spring that would result in a period of 1 s on Earth and on Mars.

Solution

(a) The period of a simple pendulum is $T = 2\pi\sqrt{L/g}$ where L is the length of the pendulum and g is the acceleration due to gravity at the pendulum's location. Thus, if a pendulum has a period of $T = 1.0$ s on Earth where $g = 9.8 \text{ m/s}^2$, its length is

$$L_{\text{Earth}} = \frac{g_{\text{Earth}} T^2}{4\pi^2} = \frac{(9.8 \text{ m/s}^2)(1.0 \text{ s})^2}{4\pi^2} = 0.25 \text{ m} = 25 \text{ cm} \qquad \Diamond$$

On Mars, where $g = 3.7 \text{ m/s}^2$, the length of a pendulum with a 1.0-s period is

$$L_{\text{Mars}} = \frac{g_{\text{Mars}} T^2}{4\pi^2} = \frac{(3.7 \text{ m/s}^2)(1.0 \text{ s})^2}{4\pi^2} = 0.094 \text{ m} = 9.4 \text{ cm} \qquad \Diamond$$

(b) The period of vibration for an object suspended from a spring is $T = 2\pi\sqrt{m/k}$ where k is the force constant of the spring and m is the mass of the object. Thus, to have a period of $T = 1.0$ s when suspended from a spring with a force constant $k = 10 \text{ N/m}$, the required mass is

$$m = \frac{kT^2}{4\pi^2} = \frac{(10 \text{ N/m})(1.0 \text{ s})^2}{4\pi^2} = 0.25 \text{ kg}$$

Both k and m are constants of the system and do not depend on the location of the system. Therefore, the same mass is needed on Earth and Mars, or

$$m_{\text{Earth}} = m_{\text{Mars}} = 0.25 \text{ kg} \qquad \Diamond$$

Note that while the same mass is needed on Earth and on Mars, the weight $w = mg$ will be different on the two planets. Thus, the oscillations occur about a different equilibrium position on Mars than on Earth.

42.   Ocean waves are traveling to the east at 4.0 m/s with a distance of 20 m between crests. With what frequency do the waves hit the front of a boat (a) when the boat is at anchor and (b) when the boat is moving westward at 1.0 m/s?

Solution

(a)   The relation between the wavelength, frequency, and wave speed is

$$v = \lambda f$$

The wavelength, λ, is the distance between successive crests of the wave, v is the speed at which the wave moves relative to the observer, and the frequency, f, is the number of crests that arrive at the observer each second.

The ocean waves have a distance $\lambda = 20$ m between crests. When the boat is at rest relative to the water, the speed of the waves relative to the boat is the same as their speed relative to the water, or $v = 4.0$ m/s. Thus, the frequency at which crests arrive at the boat is

$$f = \frac{v}{\lambda} = \frac{4.0 \text{ m/s}}{20 \text{ m}} = 0.20 \text{ s}^{-1} = 0.20 \text{ Hz} \qquad \Diamond$$

(b)   If we take the positive direction for velocities to be eastward, the velocity of the waves relative to the water is $\vec{v}_{wave,\,water} = +4.0$ m/s and the velocity of the waves relative to the boat is

$$\vec{v}_{wave,\,boat} = \vec{v}_{wave,\,water} + \vec{v}_{water,\,boat}$$

If the boat is moving westward through the water at 1.0 m/s, then $\vec{v}_{boat,\,water} = -1.0$ m/s and the velocity of the water relative to the boat is $\vec{v}_{water,\,boat} = -\left(\vec{v}_{boat,\,water}\right) = -(-1.0 \text{ m/s}) = +1.0$ m/s. Thus,

$$\vec{v}_{wave,\,boat} = +4.0 \text{ m/s} + 1.0 \text{ m/s} = +5.0 \text{ m/s}$$

The frequency at which waves hit the front of the boat in this case is

$$f = \frac{v}{\lambda} = \frac{5.0 \text{ m/s}}{20 \text{ m}} = 0.25 \text{ s}^{-1} = 0.25 \text{ Hz} \qquad \Diamond$$

47.   A simple pendulum consists of a ball of mass 5.00 kg hanging from a uniform string of mass 0.0600 kg and length L. If the period of oscillation of the pendulum is 2.00 s, determine the speed of a transverse wave in the string when the pendulum hangs vertically.

Solution

The period of a simple pendulum is $T = 2\pi\sqrt{L/g}$. Thus, if this pendulum has a period of $T = 2.00$ s, its length must be

$$L = \frac{gT^2}{4\pi^2} = \frac{(9.80 \text{ m/s}^2)(2.00 \text{ s})^2}{4\pi^2} = 0.993 \text{ m}$$

The mass per unit length of the string is then

$$\mu = \frac{m_{string}}{L} = \frac{0.0600 \text{ kg}}{0.993 \text{ m}} = 0.0604 \text{ kg/m}$$

When the pendulum hangs vertically and stationary, the tension in the string is

$$F = m_{ball}g = (5.00 \text{ kg})(9.80 \text{ m/s}^2) = 49.0 \text{ N}$$

The speed of a transverse wave in the string under these conditions will be

$$v = \sqrt{\frac{F}{\mu}} = \sqrt{\frac{49.0 \text{ N}}{0.0604 \text{ kg/m}}} = 28.5 \text{ m/s}$$

◊

50. The elastic limit of a piece of steel wire is 2.70×10^9 Pa. What is the maximum speed at which transverse wave pulses can propagate along the wire without exceeding its elastic limit? (The density of steel is 7.86×10^3 kg/m^3.)

Solution

When a wire has tension F in it, the tensile stress is $Stress = F/A$, where A is the cross-sectional area of the wire. If the mass per unit length of the wire is $\mu = m/L$, the speed of transverse waves in the wire may be written as

$$v = \sqrt{\frac{F}{\mu}} = \sqrt{\frac{A \cdot Stress}{m/L}} = \sqrt{\frac{(A \cdot L) Stress}{m}}$$

The mass of the wire is given by $m = \rho V = \rho(A \cdot L)$ where ρ is the density of the material making up the wire. The speed of transverse waves in the wire then becomes

$$v = \sqrt{\frac{(A \cdot L) Stress}{\rho(A \cdot L)}} = \sqrt{\frac{Stress}{\rho}}$$

If the maximum stress that can exist in a wire of density $\rho = 7.86 \times 10^3$ kg/m^3 is $(Stress)_{max} = 2.70 \times 10^9$ Pa, the maximum speed of transverse waves in this wire is

$$v_{max} = \sqrt{\frac{(Stress)_{max}}{\rho}} = \sqrt{\frac{2.70 \times 10^9 \text{ Pa}}{7.86 \times 10^3 \text{ kg/m}^3}} = 586 \text{ m/s} \qquad \Diamond$$

55.   A large block P executes horizontal simple harmonic motion as it slides across a frictionless surface with a frequency $f = 1.50$ Hz. Block B rests on it, as shown in Figure P13.55, and the coefficient of static friction between the two is $\mu_s = 0.600$. What maximum amplitude of oscillation can the system have if block B is not to slip?

Solution

If block B is not to slip on block P, it must always have the same velocity (and hence, rate of change in velocity or acceleration) as does block P. From the free-body diagram of block B given at the right, we have

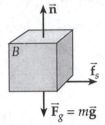

$$\Sigma F_y = ma_y = 0 \quad \text{or} \quad n = mg$$

and $\Sigma F_x = ma$ \qquad or \qquad $a = \dfrac{f_s}{m}$

Thus, the maximum acceleration block B can be given is determined by the maximum static friction force between it and block P. This is

$$\left(f_s\right)_{max} = \mu_s n = \mu_s mg$$

and the maximum acceleration of the blocks before slipping will occur is

$$a_{max} = \frac{\left(f_s\right)_{max}}{m} = \frac{\mu_s mg}{m} = \mu_s g$$

A system undergoing simple harmonic motion has its greatest acceleration at maximum displacement from equilibrium (that is, at $x = A$), and this acceleration is given by $a = A\omega^2 = A\left(2\pi f\right)^2$ where f is the frequency of vibration. Thus, the maximum amplitude of oscillation that does not cause slipping is

$$A_{max} = \frac{a_{max}}{\left(2\pi f\right)^2} = \frac{\mu_s g}{\left(2\pi f\right)^2} = \frac{(0.600)(9.80 \text{ m/s}^2)}{4\pi^2 \left(1.50 \text{ s}^{-1}\right)^2} = 6.62 \times 10^{-2} \text{ m} = 6.62 \text{ cm} \qquad \Diamond$$

63. A light balloon filled with helium of density 0.180 kg/m³ is tied to a light string of length $L = 3.00$ m. The string is tied to the ground, forming an "inverted" simple pendulum (Fig. P13.63a). If the balloon is displaced slightly from equilibrium, as in Figure P13.63b, show that the motion is simple harmonic, and determine the period of the motion. Take the density of air to be 1.29 kg/m³. (*Hint:* Use an analogy with the simple pendulum discussed in the text, and see Chapter 9.)

Solution

At equilibrium, the string of the balloon is vertical. In the sketch at the right, the string (of length L) is at angle θ from the vertical, and the balloon is displaced distance $s = L\theta$ along a circular arc from the equilibrium position. In this orientation, the net force acting tangential to the circular path followed by the balloon is

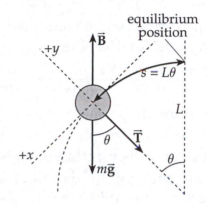

$$F_{net} = \Sigma F_x = -B \sin\theta + mg \sin\theta = -(B - mg)\sin\theta$$

The buoyant force acting on the balloon is $\quad B = \begin{pmatrix} \text{weight of} \\ \text{displaced air} \end{pmatrix} = (\rho_{air}V)g$

and the weight of the balloon is $\qquad mg = (\rho_{helium}V)g$

so the net force acting tangential to the path becomes

$$F_{net} = -\left[(\rho_{air}V)g - (\rho_{helium}V)g\right]\sin\theta = -(\rho_{air} - \rho_{helium})Vg \sin\theta$$

If the amplitude of the oscillations is small, then θ is always a small angle and we may use the approximation $\sin\theta \approx \theta = s/L$ (where θ is in radians). Then, the net force moving the balloon becomes

$$F_{net} = -(\rho_{air} - \rho_{helium})Vg \, s/L = -\left[\frac{(\rho_{air} - \rho_{helium})Vg}{L}\right]s = -k \cdot s$$

and we see that the net force is in the form of Hooke's law with an effective spring constant $\quad k = (\rho_{air} - \rho_{helium})Vg/L$

Thus, the balloon will carry out simple harmonic motion with a period given by

$$T = 2\pi\sqrt{\frac{m}{k}} = 2\pi\sqrt{\frac{\rho_{helium}V}{(\rho_{air} - \rho_{helium})Vg/L}} = 2\pi\sqrt{\left(\frac{\rho_{helium}}{\rho_{air} - \rho_{helium}}\right)\frac{L}{g}} \qquad \Diamond$$

If $L = 3.00$ m, $\rho_{helium} = 0.180$ kg/m³, and $\rho_{air} = 1.29$ kg/m³, this period is

$$T = 2\pi \sqrt{\left(\frac{0.180 \text{ kg/m}^3}{1.29 \text{ kg/m}^3 - 0.180 \text{ kg/m}^3}\right) \frac{3.00 \text{ m}}{9.80 \text{ m/s}^2}} = 1.40 \text{ s}$$ ◊

---

67.   A object of mass $m_1 = 9.0$ kg is in equilibrium while connected to a light spring of constant $k = 100$ N/m that is fastened to a wall, as in Figure P13.67a. A second object, of mass $m_2 = 7.0$ kg, is slowly pushed up against m_1, compressing the spring by the amount $A = 0.20$ m, as shown in Figure P13.67b. The system is then released, causing both objects to start moving to the right on the frictionless surface. (a) When m_1 reaches the equilibrium point, m_2 loses contact with m_1 (Fig. P13.67c) and moves to the right with velocity $\vec{v}$. Determine the magnitude of $\vec{v}$. (b) How far apart are the objects when the spring is fully stretched for the first time (Fig. P13.67d)? (**Hint:** First determine the period of oscillation and the amplitude of the m_1-spring system after m_2 loses contact with m_1.)

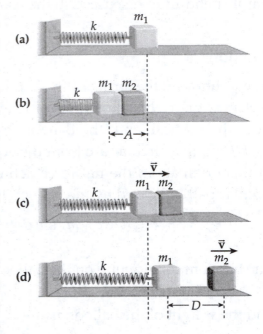

Figure P13.67

Solution

(a)   Using conservation of energy from the moment of release to the instant of separation gives

$$\left(KE + PE_g + PE_s\right)_f = \left(KE + PE_g + PE_s\right)_i$$

or $\frac{1}{2}(m_1 + m_2)v_{max}^2 + 0 + 0 = 0 + 0 + \frac{1}{2}kA^2$

Thus, $v_{max} = A\sqrt{\dfrac{k}{m_1 + m_2}} = (0.20 \text{ m})\sqrt{\dfrac{100 \text{ N/m}}{(9.0 + 7.0) \text{ kg}}} = 0.50 \text{ m/s}$ ◊

(b)   After the two blocks separate, m_1 oscillates with amplitude A' found by applying

$$\left(KE + PE_g + PE_s\right)_f = \left(KE + PE_g + PE_s\right)_i$$

to the $(m_1 + \text{spring})$ system from the moment of separation until the spring is fully stretched the first time.

$$0 + 0 + \tfrac{1}{2}kA'^2 = \tfrac{1}{2}m_1 v_{max}^2 + 0 + 0$$

$$A' = v_{max}\sqrt{\frac{m_1}{k}} = (0.50 \text{ m/s})\sqrt{\frac{9.0 \text{ kg}}{100 \text{ N/m}}} = 0.15 \text{ m}$$

The period of this oscillation is

$$T = 2\pi\sqrt{\frac{m_1}{k}} = 2\pi\sqrt{\frac{9.0 \text{ kg}}{100 \text{ N/m}}} = 1.9 \text{ s}$$

so the spring is fully stretched for the first time at $t = \tfrac{1}{4}T = 0.47$ s after separation. During this time, m_2 has moved distance $x = v_{max}t$ from the point of separation. Thus, the distance separating the two blocks at this instant is

$$D = v_{max}t - A' = (0.50 \text{ m/s})(0.47 \text{ s}) - 0.15 \text{ m} = 0.086 \text{ m} = 8.6 \text{ cm} \qquad \Diamond$$

Chapter 14

Sound

NOTES FROM SELECTED CHAPTER SECTIONS

14.1   Producing a Sound Wave

14.2   Characteristics of Sound Waves

Sound waves, which have as their source vibrating objects, are **longitudinal waves** traveling through a medium such as air. The particles of the medium oscillate back and forth along the direction in which the wave travels. This is in contrast to a transverse wave, in which the vibrations of the medium are at right angles to the direction of travel of the wave.

A sound wave traveling through air creates alternating regions of high and low molecular density and air pressure. A region of high density and air pressure is called a **compression** or **condensation**; a region of lower-than-normal density is referred to as a **rarefaction**. A sinusoidal curve can be used to represent a sound wave. *There are crests in the sinusoidal wave at the points where the sound wave has condensations, and troughs where the sound wave has rarefactions.*

14.4   Energy and Intensity of Sound Waves

The **intensity** of a wave is the rate at which sound energy flows through a unit area perpendicular to the direction of travel of the wave. The faintest sounds the human ear can detect have an intensity of about 1×10^{-12} W/m^2. This intensity is called the **threshold of hearing**. The loudest sounds the ear can tolerate, at the **threshold of pain,** have an intensity of about 1 W/m^2. The sensation of loudness is approximately logarithmic in the human ear, and the relative intensity of a sound is called the **intensity level** or **decibel level**.

14.5 Spherical and Plane Waves

The **intensity** of a spherical wave produced by a point source is proportional to the average power emitted and inversely proportional to the square of the distance from the source.

14.6 The Doppler Effect

In general the Doppler effect is experienced whenever there is relative motion between source and observer. When the source and observer are moving toward each other, the frequency heard by the observer is higher than the frequency of the source. When the source and observer are moving away from each other, the observer hears a frequency lower than the source frequency.

14.8 Standing Waves

Standing waves can be set up in a string by the superposition of two wave trains traveling in opposite directions along the string. This can occur when waves reflected from one end of the string interfere with the incident wave train, under the condition that the wavelengths (and therefore frequencies) are properly matched to the length of the string. The string has a number of natural patterns of vibration, called **normal modes**. Each normal mode has a **characteristic frequency**. The lowest of these frequencies is called the **fundamental frequency**, which together with the higher frequencies form a **harmonic series**.

The figure on the right is a schematic representation of standing waves in a string of length L. In each case the envelope represents successive positions of the string during one complete cycle. Imagine that you observe the string vibrate while it is illuminated with a strobe light. The first three normal modes are shown. The points of zero displacement are called **nodes**; the points of maximum displacement are called **antinodes**.

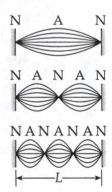

14.9 Forced Vibrations and Resonance

Consider a system (for example a mass-spring system) which has a natural frequency of vibration, f_0, and is driven or pushed back and forth with a periodic force whose frequency is f. This type of motion is referred to as a **forced vibration**. Its amplitude reaches a maximum when the frequency of the driving force equals the natural frequency of the system, f_0, called the **resonant frequency** of the system. Under this condition, the system is said to be in **resonance**.

14.10 Standing Waves in Air Columns

Sound sources can be used to produce **longitudinal standing waves** in air columns. The phase relationship between incident and reflected waves depends on whether or not the reflecting end of the air column is open or closed. This gives rise to two sets of possible standing wave conditions:

> In a **pipe open at both ends all harmonics are present;** the natural frequencies of vibration form a series with frequencies equal to integral multiples of the fundamental.

> In a **pipe closed at one end and open at the other, only odd harmonics are present.**

14.11 Beats

Consider a type of interference effect that results from the superposition of two waves with slightly different frequencies. In this situation, at some fixed point the waves are periodically in and out of phase, corresponding to an alternation in time between constructive and destructive interference. A listener hears an alternation in loudness, known as beats. The number of beats per second, or beat frequency, equals the difference in frequency between the two sources.

EQUATIONS AND CONCEPTS

A sound wave propagates through a gas or liquid as a **compressional wave**. The speed of the sound wave depends on the value of the **bulk modulus**, B (an elastic property), and the equilibrium density, ρ (an inertial property), of the material through which it is traveling.

$$v = \sqrt{\frac{B}{\rho}} \qquad (14.1)$$

$$B \equiv -\frac{\Delta P}{\Delta V/V} \qquad (14.2)$$

The **speed of sound** (or any longitudinal wave) in a solid depends on the value of **Young's modulus** and the density of the material.

$$v = \sqrt{\frac{Y}{\rho}} \qquad (14.3)$$

The **speed of sound** depends on the temperature of the medium. Equation 14.4 shows the temperature dependence in air where T is the absolute (Kelvin) temperature and 331 m/s is the speed of sound in air at 0 °C.

$$v = (331 \text{ m/s}) \sqrt{\frac{T}{273}} \qquad (14.4)$$

Speed of sound in air

The **intensity of a wave** is the rate at which energy flows across a unit area, A, in a plane perpendicular to the direction of travel of the wave. The SI units of intensity, I, are watts per square meter, W/m^2.

$$I \equiv \frac{\text{power}}{\text{area}} = \frac{\mathcal{P}}{A} \qquad (14.6)$$

At a frequency of 1 000 Hz, the faintest sound detectable by the human ear (**threshold of hearing**) has an intensity of about 10^{-12} W/m^2. An intensity of 1 W/m^2 is considered to be the greatest intensity which the ear can tolerate (**the threshold of pain**).

Human frequency range

The **decibel scale** is a logarithmic intensity scale. On this scale, the unit of sound intensity is the decibel, dB. The constant I_0 is a reference intensity, chosen to coincide with the threshold of hearing. Note: "log" indicates common logarithms (base 10) as opposed to "ln" which indicates natural logarithms (base e).

$$\beta \equiv 10 \log \left(\frac{I}{I_0} \right) \qquad (14.7)$$

$$I_0 = 1.00 \times 10^{-12} \text{ } W/m^2$$

In order to determine the decibel level corresponding to two different sources sounded simultaneously, first find the individual intensities I_1 and I_2 in W/m^2 and add these values to obtain the combined intensity $I = I_1 + I_2$. Finally, use Equation 14.7 to convert the intensity I to the decibel scale.

Calculating the decibel level due to multiple sources

The **intensity of a spherical wave produced by a point source** is inversely proportional to the square of the distance from the source. In Equation 14.8, "area" refers to the area of a spherical wave front.

$$I = \frac{\text{average power}}{\text{area}} = \frac{\mathcal{P}_{av}}{4\pi r^2} \qquad (14.8)$$

The **apparent change in frequency** heard by an observer whenever there is relative motion between the source and the observer is called the **Doppler effect**. When using Equation 14.12 the values for v_o and v_s are each substituted with either a positive or negative sign depending on the direction of their velocity. If the observer or the source moves toward the other, that velocity is substituted with a positive sign. If observer or source moves away from the other, the corresponding velocity is substituted into Equation 14.12 with a negative sign. In Equation 14.12, the velocity of sound, v, is always positive. *Also, it is important to remember that v_o (velocity of the observer) and v_s (velocity of the source) are each measured relative to the medium in which the sound travels.*

$$f_0 = f_s \left(\frac{v + v_o}{v - v_s} \right) \qquad (14.12)$$

f_s = frequency of the source
f_0 = observed frequency
v = velocity of sound (always positive)
v_o = velocity of observer
v_s = velocity of source

A series of standing wave patterns called **normal modes** can be excited in a stretched string (fixed at both ends). Each mode corresponds to a characteristic frequency and wavelength.

$$\lambda_n = \frac{2L}{n}$$

$$f_n = n f_1 = \frac{n}{2L}\sqrt{\frac{F}{\mu}} \; ; n = 1, 2, 3, \ldots \qquad (14.17)$$

Standing waves in strings

The frequency f_1 corresponding to $n = 1$, is the **fundamental frequency**, or first harmonic, and is the lowest frequency for which a standing wave is possible. The higher frequencies ($n = 2, 3, 4, \ldots$) form a harmonic series and are integral multiples of the fundamental frequency.

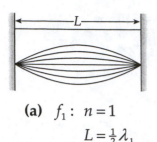

(a) f_1: $n = 1$
$L = \frac{1}{2}\lambda_1$

(b) f_2: $n = 2$
$L = \lambda_2$

(c) f_3: $n = 3$
$L = \frac{3}{2}\lambda_3$

Standing waves in a stretched string of length L fixed at both ends. The normal frequencies of vibration form a harmonic series: (a) the fundamental frequency, or first harmonic, (b) the second harmonic, and (c) the third harmonic.

In a **pipe open at both ends**, the natural frequencies of vibration form a series in which all harmonics (integer multiples of the fundamental) are present. *Note that this series of frequencies is the same as the series of frequencies produced in a string fixed at both ends.*

$$f_n = nf_1 = n\left(\frac{v}{2L}\right) \qquad (14.18)$$

where $n = 1, 2, 3, \ldots$

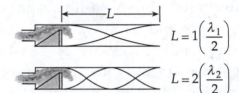

$$L = 1\left(\frac{\lambda_1}{2}\right)$$
$$L = 2\left(\frac{\lambda_2}{2}\right)$$
$$L = 3\left(\frac{\lambda_3}{2}\right)$$

In a **pipe open at one end**, only the odd harmonics (odd multiples of the fundamental) are possible.

$$f_n = n\left(\frac{v}{4L}\right) \qquad (14.19)$$

where $n = 1, 3, 5, \ldots$

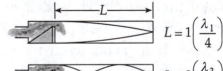

$$L = 1\left(\frac{\lambda_1}{4}\right)$$
$$L = 3\left(\frac{\lambda_3}{4}\right)$$
$$L = 5\left(\frac{\lambda_5}{4}\right)$$

SUGGESTIONS, SKILLS, AND STRATEGIES

When making calculations using Equation 14.7 which defines the intensity of a sound wave on the decibel scale, the properties of logarithms must be kept clearly in mind.

In order to determine the decibel level corresponding to two sources sounded simultaneously, you must first find the intensity, I, of each source in W/m^2, add these values, and then convert the resulting intensity to the decibel scale. As an illustration of this technique, note that if two sounds of intensity 40 dB and 45 dB are sounded together, the intensity level of the combined sources **is 46.2 dB** (**not 85 dB**).

The most likely error in using Equation 14.12 to calculate the Doppler frequency shift due to relative motion between a sound source and an observer is due to using the incorrect algebraic sign for the velocity of either the observer or the source. Remember the following relationship between relative velocity of source and observer and the corresponding Doppler frequency shift: the word **toward** is associated with an **increase** in frequency and the words **away from** are associated with a **decrease** in frequency.

REVIEW CHECKLIST

▷ Describe the harmonic displacement and pressure variation as functions of time and position for a harmonic sound wave.

▷ Calculate the speed of sound in various media in terms of appropriate elastic properties (these can include bulk modulus, Young's modulus, and the pressure-volume relationships of an ideal gas) and the corresponding inertial properties (usually the mass density).

▷ Understand the basis of the logarithmic intensity scale (decibel scale) and convert intensity values (given in W/m^2) to loudness levels on the dB scale. Determine the intensity ratio for two sound sources whose decibel levels are known. Calculate the intensity of a point source wave at a given distance from the source. Remember that you must use common logarithms (base 10) in Equation 14.7

▷ Describe the various situations under which a Doppler shifted frequency is produced. Calculate the apparent frequency for a given actual frequency for each of the various possible relative motions between source and observer.

▷ Describe in both qualitative and quantitative terms the conditions which produce standing waves in a stretched string and in an open or closed air column pipe.

SOLUTIONS TO SELECTED END-OF-CHAPTER PROBLEMS

7. You are watching a pier being constructed on the far shore of a saltwater inlet when some blasting occurs. You hear the sound in the water 4.50 s before it reaches you through the air. How wide is the inlet? (**Hint:** See Table 14.1. Assume the air temperature is 20 °C.)

Solution

From Table 14.1, the speed of sound in the saltwater is $v_w = 1530$ m/s. At $T = 20\ °C = 293$ K, the speed of sound in air is

$$v_a = (331\ \text{m/s})\sqrt{\frac{T}{273\ \text{K}}} = (331\ \text{m/s})\sqrt{\frac{293\ \text{K}}{273\ \text{K}}} = 343\ \text{m/s}$$

If d is the width of the inlet, the time required for sound traveling in air to cross the inlet is $t_a = d/v_a$. Likewise, the required transit time for sound traveling through seawater is $t_w = d/v_w$.

Since it is given that

$$t_a = t_w + 4.50\ \text{s}$$

we have

$$\frac{d}{v_a} = \frac{d}{v_w} + 4.50\ \text{s} \qquad \text{or} \qquad d = (4.50\ \text{s})\left(\frac{v_w v_a}{v_w - v_a}\right)$$

Therefore, the width of the inlet must be

$$d = (4.50\ \text{s})\left[\frac{(1530\ \text{m/s})(343\ \text{m/s})}{(1530 - 343)\ \text{m/s}}\right] = 1.99 \times 10^3\ \text{m} = 1.99\ \text{km} \qquad \Diamond$$

13.   A noisy machine in a factory produces sound with a level of 80 dB. How many identical machines could you add to the factory without exceeding the 90-dB limit?

Solution

The decibel level of a sound is given by

$$\beta = 10\log\left(\frac{I}{I_0}\right)$$

where I is the intensity of the sound wave and $I_0 = 1.0\times10^{-12}$ W/m^2 is a reference intensity. Solving for the intensity gives

$$I = I_0 10^{\beta/10}$$

The intensity of sound produced by one machine ($\beta = 80$ dB) is

$$I_1 = I_0 10^{80/10} = I_0 10^{8.0}$$

The intensity of sound necessary to reach a 90-dB level is

$$I = I_0 10^{90/10} = I_0 10^{9.0}$$

The number of machines which, acting together, would produce a 90-dB sound level is

$$N = \frac{I}{I_1} = \frac{I_0 10^{9.0}}{I_0 10^{8.0}} = 10$$

Since the factory already contains one machine, you can add 9 additional machines without exceeding the limit.   ◊

17. A train sounds its horn as it approaches an intersection. The horn can just be heard at a level of 50 dB by an observer 10 km away. (a) What is the average power generated by the horn? (b) What intensity level of the horn's sound is observed by someone waiting at an intersection 50 m from the train? Treat the horn as a point source and neglect any absorption of sound by the air.

Solution

(a)   If the train horn acts as a point source, the wave fronts are spherical with a surface area $A = 4\pi r^2$ at distance r from the horn. The average power emitted by a source can be expressed as $\mathcal{P} = IA$ where I is the intensity of the wave and A is the surface area of the wave front, both at the same distance r from the source.

The decibel level of a sound is defined as $\beta \equiv 10\log(I/I_0)$, where I is the intensity of the sound and $I_0 = 1.0 \times 10^{-12}$ W/m^2. Solving this equation for the intensity gives

$$I = I_0 \left(10^{\beta/10}\right)$$

At $r = 10$ km $= 1.0 \times 10^4$ m from the train horn, the sound level is 50 dB. Therefore, the intensity of the sound is

$$I = \left(1.0 \times 10^{-12} \text{ W/m}^2\right) 10^{50/10} = \left(1.0 \times 10^{-12} \text{ W/m}^2\right) 10^{5.0} = 1.0 \times 10^{-7} \text{ W/m}^2$$

and the average power generated by the horn must be

$$\mathcal{P} = IA = I\left(4\pi r^2\right) = \left(1.0 \times 10^{-7} \text{ W/m}^2\right)\left[4\pi\left(1.0 \times 10^4 \text{ m}\right)^2\right] = 1.3 \times 10^2 \text{ W} \qquad \Diamond$$

(b)   At a distance of $r = 50$ m, the intensity of the sound is

$$I = \frac{\mathcal{P}}{A} = \frac{\mathcal{P}}{4\pi r^2} = \frac{1.3 \times 10^2 \text{ W}}{4\pi(50 \text{ m})^2} = 4.1 \times 10^{-3} \text{ W/m}^2$$

and the decibel level is

$$\beta = 10\log\left(\frac{I}{I_0}\right) = 10\log\left(\frac{4.1 \times 10^{-3} \text{ W/m}^2}{1.0 \times 10^{-12} \text{ W/m}^2}\right) = 10\log\left(4.1 \times 10^9\right) = 96 \text{ dB} \qquad \Diamond$$

23. Two trains on separate tracks move towards one another. Train 1 has a speed of 130 km/h, train 2 a speed of 90.0 km/h. Train 2 blows its horn, emitting a frequency of 500 Hz. What is the frequency heard by the engineer on train 1?

Solution

When a source moving with velocity v_S emits sound having frequency f_S, the frequency f_O detected by an observer moving with velocity v_O is

$$f_O = f_S \left(\frac{v + v_O}{v - v_S} \right)$$

Here, v is the velocity of sound in the propagating medium. Also, v_S is positive if the source moves toward the observer and negative if the source moves away from the observer. Similarly, v_O is positive when the observer moves toward the source and negative if the observer moves away from the source.

In this case, train 2 is the source of a sound having frequency $f_S = 500$ Hz and it is moving toward the observer (train 1) at 90 km/h. Thus, $v_S = +90$ km/h. The observer is on train 1 and moves toward the source (train 2) at 130 km/h, so $v_O = +130$ km/h. The sound travels through air with a speed of

$$v = \left(345 \ \frac{m}{s} \right) \left(\frac{1 \ km}{10^3 \ m} \right) \left(\frac{3\,600 \ s}{1 \ h} \right) = 1\,240 \ km/h$$

The frequency detected by the observer is then

$$f_O = (500 \ Hz) \left(\frac{1\,240 \ km/h + 130 \ km/h}{1\,240 \ km/h - 90.0 \ km/h} \right) = 595 \ Hz$$

◊

31. The ship in Figure P14.31 travels along a straight line parallel to, and 600 m from, the shore. The ship's radio receives simultaneous signals of the same frequency from antennas A and B. The signals interfere constructively at point C, which is equidistant from A and B. The signal goes through the first minimum at point D. Determine the wavelength of the radio waves.

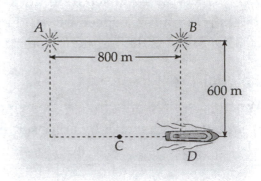

Figure P14.31

Solution

When the ship is at point D, the distance to the ship from the antenna at point A is given by the Pythagorean theorem as

$$d_A = \sqrt{(800 \text{ m})^2 + (600 \text{ m})^2} = 1\,000 \text{ m}$$

The distance to the ship from the antenna at point B is 600 m. Thus, the difference in the path lengths the radio waves must travel from the two antennas to the ship is

$$\Delta d = d_A - d_B = 1\,000 \text{ m} - 600 \text{ m} = 400 \text{ m}$$

If the two signals reaching the ship are to interfere destructively and yield a minimum of intensity, the difference in path lengths must be an odd number of half-wavelengths, or $\Delta d = \left(n + \tfrac{1}{2}\right)\lambda$ where $(n = 0, 1, 2, 3, \ldots)$. Since the ship is located at the position of the first minimum in the interference pattern produced by the waves from the two sources, n must have its lowest possible value, or $n = 0$. Hence, $\Delta d = \lambda/2$ or the wavelength of the radio waves must be

$$\lambda = 2\Delta d = 2(400 \text{ m}) = 800 \text{ m} \qquad \Diamond$$

37. Two speakers are driven by a common oscillator at 800 Hz and face each other at a distance of 1.25 m. Locate the points along a line joining the speakers where relative minima of the amplitude of the pressure would be expected. (Use $v = 343$ m/s.)

Solution

Since the speakers are driven by a common oscillator, they must vibrate in phase with each other. Thus, the point halfway between them (being equidistant from the two sources) must be an anti-node in any standing wave pattern formed.

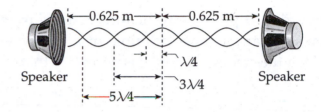

Relative minima (i.e., nodes) are located on either side of this central anti-node at a distance of

$$\frac{\lambda}{4} = \frac{1}{4}\left(\frac{v}{f}\right) = \frac{343 \text{ m/s}}{4(800 \text{ Hz})} = 0.107 \text{ m} \text{ from the midpoint.}$$

This means that they are located at distances of

$$d = 0.625 \text{ m} \pm 0.107 \text{ m} = 0.518 \text{ m} \text{ and } 0.732 \text{ m from either speaker} \qquad \lozenge$$

A second pair of nodes will be found at a distance of $3\lambda/4 = 3(0.107 \text{ m})$ on either side of the midpoint. The positions of these nodes are then located at distances of

$$d = 0.625 \text{ m} \pm 3(0.107 \text{ m}) = 0.303 \text{ m} \text{ and } 0.947 \text{ m} \text{ from either speaker} \qquad \lozenge$$

Finally a pair of nodes will be found at a distance of $5\lambda/4 = 5(0.107 \text{ m})$ on either side of the midpoint. These points are located at distances of

$$d = 0.625 \text{ m} \pm 5(0.107 \text{ m}) = 0.089 \text{ m} \text{ and } 1.16 \text{ m} \text{ from either speaker} \qquad \lozenge$$

41.   A 60.00-cm guitar string under a tension of 50.000 N has a mass per unit length of 0.100 00 g/cm. What is the highest resonant frequency that can be heard by a person capable of hearing frequencies up to 20 000 Hz?

Solution

In order to produce resonance in a string fixed at both ends, the length of the string must be any integer multiple of a half-wavelength of the transverse waves traveling along the string. That is, it is necessary that

$$L = n\left(\frac{\lambda_n}{2}\right) \quad \text{or} \quad \lambda_n = \frac{2L}{n} \text{ where } n \text{ is any positive integer}$$

Thus, the resonant frequencies of the string are $f_n = \frac{v}{\lambda_n} = n\left(\frac{v}{2L}\right)$ where v is the speed of transverse waves in the string. If a string having mass per unit length

$$\mu = \left(0.100\ 00\ \frac{g}{cm}\right)\left(\frac{10^2\ cm}{1\ m}\right)\left(\frac{1\ kg}{10^3\ g}\right) = 1.000\ 0\times10^{-2}\ kg/m$$

is under a tension of $F = 50.000$ N, the speed of transverse waves in it is

$$v = \sqrt{\frac{F}{\mu}} = \sqrt{\frac{50.000\ N}{1.000\ 0\times10^{-2}\ kg/m}} = 70.711\ m/s$$

If the length of the string is $L = 60.00$ cm, the highest resonance with frequency less than or equal to 20 000 Hz (and therefore audible to the listener) is

$$n \leq f\left(\frac{2L}{v}\right) = (20\ 000\ s^{-1})\left[\frac{2(0.6000\ m)}{70.711\ m/s}\right] = 339.41$$

Since n must have an integer value, we have $n = 339$

and the frequency at which this resonance occurs is

$$f_{339} = n\left(\frac{v}{2L}\right) = 339\left(\frac{70.711\ m/s}{2(0.600\ 0\ m)}\right) = 19\ 976\ Hz = 19.976\ kHz \qquad \Diamond$$

47.   A pipe open at both ends has a fundamental frequency of 300 Hz when the temperature is 0 °C. (a) What is the length of the pipe? (b) What is the fundamental frequency at a temperature of 30 °C?

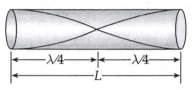

Solution

A standing wave pattern in a pipe open at both ends must have an anti-node at each end. In the fundamental mode of vibration (that is, for the lowest frequency and hence longest wavelength capable of producing resonance), the standing wave pattern is as shown in the sketch.

Therefore, the length of the pipe is $L = \lambda/4 + \lambda/4 = \lambda/2$

(a)   The speed of sound in air at absolute temperature T is

$$v = (331 \text{ m/s})\sqrt{T/273 \text{ K}}$$

Thus, at $T = 0 \text{ °C} = 273 \text{ K}$, $v = 331 \text{ m/s}$

If the fundamental frequency of the pipe is 300 Hz,

the length of the pipe is $L = \dfrac{\lambda}{2} = \dfrac{v}{2f} = \dfrac{331 \text{ m/s}}{2(300 \text{ Hz})} = 0.552 \text{ m}$ ◊

(b)   At $T = 30 \text{ °C} = 303 \text{ K}$

the speed of sound is $v = (331 \text{ m/s})\sqrt{\dfrac{303 \text{ K}}{273 \text{ K}}} = 349 \text{ m/s}$

The change in the length of the pipe as the temperature rises from 0 °C to 30 °C is negligible in comparison to the total length. Thus, the length of the pipe is still $L = 0.552 \text{ m}$ and, in the fundamental resonance mode, $\lambda = 2L$. The fundamental frequency at this temperature is then

$$f = \dfrac{v}{\lambda} = \dfrac{349 \text{ m/s}}{2(0.552 \text{ m})} = 316 \text{ Hz}$$ ◊

53.   A student holds a tuning fork oscillating at 256 Hz. He walks toward a wall at a constant speed of 1.33 m/s. (a) What beat frequency does he observe between the tuning fork and its echo? (b) How fast must he walk away from the wall to observe a beat frequency of 5.00 Hz?

Solution

Initially, the student serves as a moving source $(v_S = v_{student})$ of sound having frequency $f_S = 256$ Hz. The wall is a stationary $(v_O = 0)$ receiver or observer of this sound, so the sound wave received and reflected by the wall has frequency

$$f_{echo} = f_O' = f_S \left(\frac{v + v_O}{v - v_S} \right) = f_S \left(\frac{v}{v - v_{student}} \right)$$

where $v = 345$ m/s is the speed of sound in air

At the reflection, the wall serves as a stationary source $(v_S = 0)$ of sound having frequency f_{echo}, and the student is a moving observer $(v_O = v_{student})$ of this sound. The frequency of the echo heard by the student is

$$f_O = f_{echo} \left(\frac{v + v_O}{v - v_S} \right) = \left[f_S \left(\frac{v}{v - v_{student}} \right) \right] \left(\frac{v + v_{student}}{v} \right) = f_S \left(\frac{v + v_{student}}{v - v_{student}} \right)$$

The beat frequency heard by the student is then

$$f_{beat} = |f_O - f_S| = \left| f_S \left(\frac{v + v_{student}}{v - v_{student}} \right) - f_S \right| = 2 f_S \left| \frac{v_{student}}{v - v_{student}} \right|$$

(a)   If the student *approaches* the wall at speed $|v_{student}| = 1.33$ m/s, the needed velocity is $v_{student} = +1.33$ m/s, and the beat frequency is

$$f_{beat} = 2 f_S \left| \frac{v_{student}}{v - v_{student}} \right| = 2(256 \text{ Hz}) \left| \frac{+1.33 \text{ m/s}}{345 \text{ m/s} - 1.33 \text{ m/s}} \right| = 1.98 \text{ Hz} \qquad ◊$$

(b)   If the student *moves away* from the wall at speed $|v_{student}|$, the needed velocity is $v_{student} = -|v_{student}|$. If the student is to hear $f_{beat} = 5.00$ Hz, then

$$2 f_S \left| \frac{-|v_{student}|}{v + |v_{student}|} \right| = 5.00 \text{ Hz} \qquad \text{or} \qquad \frac{2 f_S |v_{student}|}{v + |v_{student}|} = 5.00 \text{ Hz}$$

This yields

$$|v_{student}| = v \left(\frac{5.00 \text{ Hz}}{2 f_S - 5.00 \text{ Hz}} \right) = \left(345 \frac{\text{m}}{\text{s}} \right) \left[\frac{5.00 \text{ Hz}}{2(256 \text{ Hz}) - 5.00 \text{ Hz}} \right] = 3.40 \text{ m/s} \qquad ◊$$

59. On a workday the average decibel level of a busy street is 70 dB, with 100 cars passing a given point every minute. If the number of cars is reduced to 25 every minute on a weekend, what is the decibel level of the street?

Solution

The decibel level of a sound is given by $\beta = 10\log\left(\dfrac{I}{I_0}\right)$

where I is the intensity of the sound and I_0 is a reference intensity.

On the weekend, there are one-fourth as many cars passing per minute as on a week day. Thus, the expected sound intensity, I_2, on the weekend should be one-fourth the sound intensity, I_1, on a week day. The difference in the decibel levels on a week day and on the weekend will be

$$\beta_1 - \beta_2 = 10\log\left(\frac{I_1}{I_0}\right) - 10\log\left(\frac{I_2}{I_0}\right) = 10\log\left(\frac{I_1}{I_2}\right) = 10\log(4) = 6 \text{ dB}$$

The decibel level on the weekend is

$$\beta_2 = \beta_1 - 6 \text{ dB} = 70 \text{ dB} - 6 \text{ dB} = 64 \text{ dB}$$

◊

66. Two identical speakers separated by 10.0 m are driven by the same oscillator with a frequency of $f = 21.5$ Hz (Fig. P14.66). Explain why a receiver at A records a minimum in sound intensity from the two speakers. (b) If the receiver is moved in the plane of the speakers, what path should it take so that the intensity remains at a minimum? That is, determine the relationship between x and y (the coordinates of the receiver) such that the receiver will record a minimum in sound intensity. Take the speed of sound to be 344 m/s.

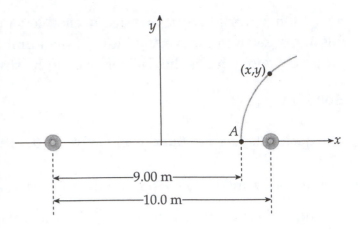

Figure P14.66

Solution

(a)   The speakers generate sound waves having a wavelength λ, where

$$\lambda = \frac{v}{f} = \frac{344 \text{ m/s}}{21.5 \text{ Hz}} = 16.0 \text{ m}$$

The difference in the path lengths from the two speakers to point A is

$$\Delta d = d_1 - d_2 = 9.00 \text{ m} - 1.00 \text{ m} = 8.00 \text{ m} = \lambda/2$$

Since this difference in path lengths is an odd multiple of a half-wavelength, the waves from the two speakers arrive at point A out of phase with each other, producing destructive interference and a minimum in intensity   ◇

(b)   The receiver should move so the difference in path lengths remains equal to $\Delta d = d_1 - d_2 = \lambda/2$. Thus, using the Pythagorean theorem and the triangles shown in the sketch at the right,

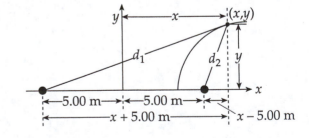

$$\sqrt{(x+5.00 \text{ m})^2 + y^2} - \sqrt{(x-5.00 \text{ m})^2 + y^2} = 8.00 \text{ m}$$

or

$$\sqrt{(x+5.00 \text{ m})^2 + y^2} = \sqrt{(x-5.00 \text{ m})^2 + y^2} + 8.00 \text{ m}$$

Squaring this equation and simplifying yields

$$\frac{5x}{4} - 4.00 \text{ m} = \sqrt{(x-5.00 \text{ m})^2 + y^2}$$

Again, we square this equation and simplify to obtain $9x^2 - 16y^2 = 144 \text{ m}^2$

or $\qquad \dfrac{x^2}{16} - \dfrac{y^2}{9} = 1.00 \text{ m}^2 \qquad$ which is the equation of a hyperbola. $\qquad \Diamond$

---

69. Two ships are moving along a line due east. The trailing vessel has a speed of 64.0 km/h relative to a land-based observation point, and the leading ship has a speed of 45.0 km/h relative to the same station. The trailing ship transmits a sonar signal at a frequency of 1 200 Hz. What frequency is monitored by the leading ship? (Use 1 520 m/s as the speed of sound in ocean water.)

Solution

The source of the sound waves (the trailing ship) is *moving toward* the observer (leading ship). Thus, $v_s = +64.0$ km/h

The observer (leading ship) is *moving away* from the source (trailing ship), so $v_O = -45.0$ km/h. The sonar waves are transmitted at a frequency $f_s = 1\,200$ Hz and travel through the propagating medium (sea water) at

$$v = \left(1\,520 \ \frac{\text{m}}{\text{s}}\right)\left(\frac{1 \text{ km}}{10^3 \text{ m}}\right)\left(\frac{3\,600 \text{ s}}{1 \text{ h}}\right) = 5\,470 \text{ km/h}$$

Therefore, the frequency detected by the leading ship will be

$$f_O = f_s\left(\frac{v + v_O}{v - v_s}\right) = (1\,200 \text{ Hz})\left(\frac{5\,470 \text{ km/h} - 45.0 \text{ km/h}}{5\,470 \text{ km/h} - 64.0 \text{ km/h}}\right) = 1\,204 \text{ Hz} \qquad \Diamond$$